SpringerBriefs in Social Work

SpringerBriefs present concise summaries of cutting-edge research and practical applications. Featuring compact volumes of 55 to 125 pages, the series covers a range of content from professional to academic. Typical topics might include:

- A timely report of state-of-the art analytical techniques
- A bridge between new research results, as published in journal articles, and a contextual literature review
- A snapshot of a hot or emerging topic
- An in-depth case study or clinical example
- A presentation of core concepts that students must understand in order to make independent contributions

SpringerBriefs in Social Work showcases emerging theory, empirical research, and practical applications in a wide variety of topics in social work and related areas. Briefs are characterized by fast, global electronic dissemination and expedited production schedules.

Stephanie Rost

Climate Resilience Guide for Social Workers

Springer

Stephanie Rost
Department of Social Work
University of Gothenburg
Gothenburg, Sweden

ISSN 2195-3104 ISSN 2195-3112 (electronic)
SpringerBriefs in Social Work
ISBN 978-3-032-24603-5 ISBN 978-3-032-24604-2 (eBook)
https://doi.org/10.1007/978-3-032-24604-2

This Springer imprint is published by the registered company Springer Nature Switzerland AG
The registered company address is: Gewerbestrasse 11, 6330 Cham, Switzerland

For my children, Jack, Nash, and Josh—you give me reason to strive for a safer future and fill my present with gratitude for being your mother.

Competing Interests The author declares no competing interests in the publication of this book.

Contents

About the Author

Stephanie Rost is a PhD candidate at Gothenburg University (Sweden), where her research focuses on climate change adaptations and their implications for social work. She has a professional background as a social worker and social work manager in Scotland, primarily in the area of child protection. Stephanie has contributed to the field through numerous peer-reviewed publications addressing these subjects.

Chapter 1
Introduction: Why Is Climate Change a Social Work Issue?

If you have picked up this book and opened it, it means something about where you are in your journey to knowledge about climate change. It suggests already that you look outside your window, you watch the news, or you are attentive to the massive changes that are unfurling before our eyes around the globe. It may mean you have a sense of alarm and are looking for tools to manage this in your personal life, in your role as a social worker, social work student, or educator. It's fair to say that thinking about it too much can be disconcerting to say the least, with horror and panic not unjustified. As individuals we feel very small in this crisis, powerless and quite alone. Not everyone is vocal in talking about it and so we are left with our own spiraling thoughts, and confusion as to why others aren't standing up and shouting loudly about our collective predicament. Why we aren't is another story.

This story is about social work, and why climate change and its impacts are relevant to our practice. For some this might be obvious, while for others they might not want to hear about a social problem outwith their field of interest. And this is where climate change breaks with our traditions in the field—every issue is important, every issue affects lives—but this issue is a matter of survival for all of us, clients and social workers alike. Our power in this is somewhat neutralized, our skills in risk assessment, support, and protection become almost meaningless in a global disaster encompassing all of humanity. That may be so, but it does not mean that we have to sit by and do nothing, stay at home and wait for the end times. No. This is not one of those scenarios, and even in the worst disaster movie, I don't notice anyone doing that. Humans have endured much, and while this is undoubtedly unique in its scale and likely consequences, we still have agency over what we do with that knowledge, where we put our energies, and who we support in this crisis.

Those who go into social work usually do so not for the salary, but from a desire at whatever level to help others—to be of use, to serve, to place the energies they have in this life to something that has a purpose. That purpose being to raise up those who are struggling, be their champion, protect the vulnerable, and keep them from harm. In the coming years where climate change impacts become more pronounced

S. Rost, *Climate Resilience Guide for Social Workers*,
SpringerBriefs in Social Work, https://doi.org/10.1007/978-3-032-24604-2_1

and risk ravaging communities, exposing vulnerabilities, and exerting considerable suffering, I ask that social workers hold on to that purpose. Because it is one that exists in each of us. And it is one that we can practice, even if the proverbial shit hits the climate fan.

Let's pause for just a second—why is this even relevant? How are climate change and social work connected? Well, unless you've been living under an air-conditioned bridge, chances are you've at least *heard* of climate change. Maybe it hasn't touched your life directly—yet. Or maybe you're out protesting every week, demanding action. Either way, the crisis isn't going away. It's growing, and its impacts are starting to seep into every part of life, including the work of social workers. Global temperatures have increased approximately at time of writing (a very important caveat these days given the speed of warming) by 1.5 degrees (NASA, n.d). That was the target that was proclaimed as the limit beyond which catastrophe would unfold with each move further up the temperature gradient (Intergovernmental panel on Climate Change, 2018). But how can 1.5 degrees mean so much? We have temperature increases of 10 degrees, 20 degrees, and 30 degrees over seasons. Why does this figure mean so much?

Well the simple complicated answer is that it is the global mean temperature. The average temperature. Similarly to how our body holds a temperature of 37 degrees, the planet has been holding its temperature at a steady 14 degrees on average for the last 12,000 years or so (Kaufman et al., 2020). That timeframe is known as the Holocene and it is within this that humans have taken off in their advancement, as we had stable seasons and predictability around them. We could plant crops, yes there would be extremes, but even these were predictable to a degree. We have temperate zones, monsoons in a regular occurrence, a hurricane "season." All these allow us to prepare for potential disasters and adapt accordingly. We are moving out of that zone, into one that humans have not experienced before (Rockström, 2017). With this, comes the risk of seasons losing that predictability, more extreme weather making it more difficult to grow crops, water availability reducing due to heat, temperatures in some areas becoming so high they make places unlivable (IPCC, 2022; Lesk et al., 2016; Stanford University, 2021; Xu et al., 2020). In blatant terms, we are making the planet, our home, inhospitable to humans.

The United Nations Security Council, the U.S. military and the Intergovernmental panel on climate change all agree on one thing about climate change. It will exacerbate existing vulnerabilities, making them worse (CNA Military Advisory Board, 2007; IPCC, 2023; UN Security Council, 2019). Vulnerability is what we as social workers deal with on a daily basis, assessing it, supporting those experiencing it or analyzing the structures that create it. Vulnerability as horrible as it sounds is our bread and butter. We know it, we think we understand it and have a grasp of how it affects individuals, groups and communities. Climate change not only makes those same groups more vulnerable but creates new groups and additional vulnerabilities that we had not considered (World Bank, n.d).

Women as an example—collectively climate change does not do anything to improve the lot of women. Incidents of domestic violence related to climate change and heat have already been seen as increasing (Koubi & Schaub, 2024). Among older people, it is older women that are assessed as being more affected by the increasing temperatures through physical health consequences (Leach et al., 2024). This is the reason many courageous women in Switzerland took their government to court arguing that their human right to health had been breached, and won (European Court of Human Rights, 2024). For women in countries with severe food shortages and hunger, women are often the hardest hit (Concern Worldwide, 2024). We know that globally around 47.8 million more women than men experience food insecurity (UN women, 2024). Given that food insecurity is exacerbated by climate changes, women face further hardship in this regard.

Children, often the focus of social work intervention due to their vulnerability engendered by their age, have proven to be also vulnerable to climate changes and its impacts. From heat stress affecting children more, we read of them collapsing on walks with their parents who had unwittingly exposed their child to perilous heat conditions (Communications Earth & Environment, 2024). Pollution in the atmosphere affects the lungs of children more, leading to respiratory illness (European Environment Agency, 2022). Beyond physical health, the emotional health of children and young people demonstrates increasing amounts of anxiety and a sense of despair about their futures (Hickman et al., 2021). With just cause.

But perhaps one of the greatest injustices of climate change is how it punishes those least responsible. The communities that contribute the least to greenhouse gas emissions are often the ones who suffer first and hardest (Roser & Ritchie, 2023). Low-income populations, Indigenous communities, and those in the Global South face disproportionate exposure to environmental hazards, with fewer resources to adapt or recover (Pereira & Silva, 2023). Climate change intensifies existing structural inequalities, compounding generations of injustice (Oxfam, n.d). In the words of an inspirational speaker to me, “People are not being left behind, they are being pushed back” (Njehû, 2025). This phrase highlights the close relationship between the economic systems in which we live, climate change impacts developed through overconsumption by the global north, and the further insult to injury of asking the poorest of the world to pay the biggest price for the consequences.

For social workers committed to social justice, this represents a fundamental moral and professional challenge. We cannot address inequality without addressing climate change, and we cannot address climate change without confronting inequality. These are but a few of those groups affected, but I think they show an indication of how social work will likely need to be directly involved with those who are impacted by climate changes. Because those in need are those who we already work with, but in greater numbers. This risks overwhelming already strained social work organizations, with many at risk of falling through the cracks.

I remember when the reality of climate change truly hit me. I was on maternity leave with my third child, watching a video by Mark Lynas about the consequences of each degree of warming. As the temperature climbed with his descriptions, so did a deep, rising fear-for my children, for all children. That was the moment it all clicked: this isn’t just an environmental issue, it’s a profound injustice. We’ve built this crisis, and now we owe it to the next generation to face it with honesty and action. Social workers have the same goal, to try to take away the structures that block, to give assistance where needed and protect where harm is inflicted. We now must join the dots and work together.

Social work is ethically bound not to look away, not only as part of our profession but as our full ethical selves, who are in a sense called to this profession much as doctors or nurses may be—from a desire to help do something about the pain we see. This guide does not promise easy answers, but it does provide direction. It addresses the hard questions raised by climate change and its societal impacts, offering practical advice and action points. Drawing from diverse fields within social work, it brings together knowledge, case examples, and critical reflections to help practitioners prepare, adapt, and support others in an era of intensifying crisis.

References

Communications Earth & Environment. (2024). Heat exposure, injuries and vulnerability among preschool-aged children during heat waves. *Communications Earth & Environment.* Advance online publication. https://doi.org/10.1038/s43247-024-01846-1

Concern Worldwide. (2024, April 11). *Explainer: Five ways climate change hits women the hardest* [News release]. Concern Worldwide. Retrieved June 26, 2025, from https://www.concern.net/news/women-and-climate-change-gender-inequality

CNA Military Advisory Board. (2007). *National security and the threat of climate change.* CNA Corporation. https://www.cna.org/reports/2007/national-security-and-the-threat-of-climate-change

European Court of Human Rights. (2024, April 9). *Verein KlimaSeniorinnen Schweiz and others v. Switzerland*, App. No. 53600/20, Grand Chamber, [2024] ECHR 304.

European Environment Agency. (2022, June). *Air pollution and children's health.* EEA. Retrieved June 26, 2025, from https://www.eea.europa.eu/publications/air-pollution-and-childrens-health

Hickman, C., Marks, E., Pihkala, P., Clayton, S., Lewandowski, R. E., Mayall, E., & van Susteren, L. (2021). Climate anxiety in children and young people and their beliefs about government responses to climate change: A global survey. *The Lancet Planetary Health, 5*(12), e779–e786.

Intergovernmental Panel on Climate Change. (2018). *Global warming of 1.5 °C: An IPCC Special Report on the impacts of global warming of 1.5 °C above pre-industrial levels and related global greenhouse gas emission pathways.* IPCC.

IPCC. (2022). *Climate change 2022: Impacts, adaptation and vulnerability. Contribution of Working Group II to the Sixth Assessment Report of the Intergovernmental Panel on Climate Change* (H.-O. Pörtner, D. C. Roberts, M. Tignor, E. Poloczanska, K. Mintenbeck, A. Alegría, ... & B. Rama, Eds.). Cambridge University Press. https://www.ipcc.ch/report/ar6/wg2

Intergovernmental Panel on Climate Change (IPCC). (2023). *Climate change 2023: Synthesis report. Summary for policymakers.* https://www.ipcc.ch/report/sixth-assessment-report-synthesis-report/

Kaufman, D. S., McKay, N. P., Routson, C. C., Erb, M. P., Zander, P. D., Boyd, M., & Hakim, G. J. (2020). A global database of Holocene paleotemperature records. *Scientific Data, 7*(1), 115. https://doi.org/10.1038/s41597-020-0445-3

Koubi, V., & Schaub, M. (2024). Climate shocks and intimate partner violence: Evidence from India. *PLOS Climate, 3*(2), Article e0000183. https://doi.org/10.1371/journal.pclm.0000183

Leach, O. K., Cottle, R. M., Fisher, K. G., Wolf, S. T., & Kenney, W. L. (2024). Sex differences in heat stress vulnerability among middle-aged and older adults (PSU HEAT Project). *American Journal of Physiology-Regulatory, Integrative and Comparative Physiology, 327*(3), R320–R327. https://doi.org/10.1152/ajpregu.00114.2024

Lesk, C., Rowhani, P., & Ramankutty, N. (2016). Influence of extreme weather disasters on global crop production. *Nature, 529*(7584), 84–87. https://doi.org/10.1038/nature16467/

NASA. (n.d.). *Global temperature* [Vital Signs of the Planet]. NASA. Retrieved June 26, 2025, from https://climate.nasa.gov/vital-signs/global-temperature/?intent=121

Njehû, N. N. (2025, June). *Confronting inequalities in the green transformation* [Conference presentation]. 18th Conference of the International Society for Ecological Economics & 11th International Degrowth Conference, Oslo, Norway.

Oxfam. (n.d.). *Climate change and inequality*. Oxfam. Retrieved June 26, 2025, from https://www.oxfamamerica.org/explore/issues/climate-action/climate-change-and-inequality/

Pereira, L. C., & Silva, R. M. (2023). Climate change impacts on vulnerable communities in the Global South: A systematic review. *Environmental Development, 45*, Article 100803. https://doi.org/10.1016/j.envdev.2022.100803

Rockström, J. (2017, September 26). *Presenting a framework for preserving Earth's resilience* [Lecture transcript]. MIT Environmental Solutions Initiative. Retrieved June 26, 2025, from MIT News website.

Roser, M., & Ritchie, H. (2023). *Global inequalities in CO_2 emissions*. Our World in Data. https://ourworldindata.org/inequality-co2

Stanford University. (2021, June 3). *Warming world may threaten weather predictability*. Stanford News. https://news.stanford.edu/2021/06/03/warming-world-may-threaten-weather-predictability/

United Nations Security Council. (2019, January 25). *Security Council debates link between climate change and conflict*. United Nations Meetings Coverage. https://press.un.org/en/2019/sc13677.doc.htm

UN Women. (2024). *The gender snapshot 2024: SDG 2 – Zero hunger*. UN Women. Retrieved June 26, 2025, from https://www.unwomen.org/en/resources/gender-snapshot/sdg-2

World Bank. (n.d.). *Social dimensions of climate change*. Retrieved June 26, 2025, from https://www.worldbank.org/en/topic/social-dimensions-of-climate-change

Xu, C., Kohler, T. A., Lenton, T. M., Svenning, J.-C., & Scheffer, M. (2020). Future of the human climate niche. *Proceedings of the National Academy of Sciences, 117*(21), 11350–11355. https://doi.org/10.1073/pnas.1910114117

Chapter 2
The Climate Future Humanity Faces

Introduction

So, what are we facing? There is no sugar coating what I am about to write. There is no upside, no spin I can put on the factual data of what climate changes are and will continue to bring. This is not easy to write, as I too live on the same blue speck in a glimmer of light that you do, so to face the fundamental truths of the crisis is not a casual or intellectual exercise. It is marked by human emotion, sadness, and yes, trepidation.

The earth has experienced climatic changes throughout its lifespan. These have included extremely high temperatures and plunging lows (NASA, 2020). There have been countless species that have come and gone within the lifetime of the earth, the planet the only witness to those that evolved, thrived, and then became extinct. What we know of our own species is that we have evolved to live and thrive within a narrow temperature window. Above a wet bulb temperature ranging from 22 to 35 degrees (Sherwood & Huber, 2010), the human body cannot cool itself, and will die. The ancient temperature spikes which are often quoted as examples of the idea that the "climate has always changed," are in fact true. What is also true is that human beings would not have been able to survive in them, and we evolved to survive in much lower temperatures (Potts, 1998). Additionally, for the last 12,000 years or so, we have been able to develop agriculture, due to stable seasons and predictable weather (Richerson & Boyd, 2005). Prior to this, our populations were low, as homosapiens endured through ice ages into this inter glacial period. The Holocene period of time has allowed everything we see around us to be developed—human civilization and proliferation of humans throughout the globe (Lenton & Williams, 2013).

The reason I give this backdrop is that we are now at a point where there are approximately 9 billion humans on the planet. And that is where we begin.

Global mean surface temperature has been a comfortable average of 14 degrees Celsius (Kaufman et al., 2022). It has now increased to approximately 1.5 degrees Celsius above pre-industrial levels, before we began emitting carbon dioxide through

S. Rost, *Climate Resilience Guide for Social Workers*,
SpringerBriefs in Social Work, https://doi.org/10.1007/978-3-032-24604-2_2

industrialized processes (WMO, 2024). This had appeared to be increasing in a linear fashion in direct response to the amount of carbon dioxide we were emitting. This allowed estimates to be developed of the future temperature increases, and reports such as the Intergovernmental panel on climate change reports summarizing these findings. What is not commonly known is that in their temperature scenarios, they included the likely future development of carbon capture and storage at a large scale, so predicting that the temperatures wouldn't reach certain heights because we would think of something to fix it. This was actually included in their estimates (IPCC, 2022). What wasn't included in their estimates was the idea of crossing thresholds of temperature that lead to tipping points in the earth systems or unknown variables that might interact with one another, causing heating to accelerate (IPCC, 2023). And this is unfortunately what appears to have happened (Hansen et al., 2023). The earth sensitivity to carbon dioxide and other greenhouse gases appears to be higher than originally predicted, and those earlier scenarios of future temperatures are now drawn forward (ibid). Meaning, that it's likely to get hotter faster. And as this happens, there is then even greater risk of passing further thresholds, (IPCC, 2023) pushing the earth's climate into one that we may not be able to survive in. At least not in the numbers of humans that exist today, and not without significant adaptation to our physical environments and infrastructure, to which there has been extremely limited change, as well as continued increasing emissions from fossil fuels (Greenfacts, n.d).

The future projected, is one based on climate science, and accepts that there is a *chance* that governments will unite to eliminate fossil fuel use. I leave it to you as to whether you believe that is likely. Thirty years of climate conferences and ever-increasing emissions would appear to suggest not. In this future, climate extremes are predicted and have been proven to increase (IPCC, 2021). This means increasing storms, hurricanes, extreme floods, wildfires and droughts. These do not occur in isolation but have compounding effects on agriculture, housing, livelihoods, and water supplies as well as the secondary social consequences such conflict, ill health and disease, homelessness, and deprivation (IPCC, 2023).

Based on the latest information regarding temperature increases and including accelerating temperatures, Table 2.1 depicts the expected temperatures by year and indicates that global warming has accelerated. Adapted from a table by Simons (2025), with formatting modified for clarity. The underlying data are from Foster and Rahmstorf (2025).

As you might be surprised to see, these are beyond what has traditionally been expected or even explored in adaptation planning. 1.5 degrees was the theoretical limit past which humans were described as entering dangerous unknown territory, 2-degree projections were described as catastrophic (IPCC, 2018). At the point of writing the likelihood appears to be that we will sail right past 2 degrees on our way to 3 degrees, then 4 and beyond. Beyond 4 or 5 degrees, the possibility of humanity surviving becomes low, societal collapse on a global scale is likely, mass famines and billions of deaths due to heat, hunger, conflict, and weather extremes (Lynas, 2007). And on that chart, the timelines for these things are not far away—they are within the lifetimes of those reading this—and absolutely of your children.

Table 2.1 Global warming has accelerated significantly

Data	Ending value (°C)	Rate (°C/ decade)	Cross + 1.5 °C	Cross + 2.0 °C	Cross + 2.5 °C	Cross + 3.0 °C	Cross + 3.5 °C	Cross + 4.0 °C
NASA	1,45	0,42	2026	2037	2049	2061	2073	2085
NOAA	1,45	0,42	2026	2037	2049	2061	2073	2085
HadCRU	1,42	0,39	2026	2039	2052	2065	2077	2090
Berkeley	1,45	0,43	2026	2037	2048	2060	2072	2083
ERA5	1,54	0,48	2024	2034	2044	2054	2065	2075
Average	1,46	0,43	2026	2037	2048	2060	2072	2084

Sources Reprinted with permission from Simons (2025), Foster and Rahmstorf (2025)[1]

But rising temperatures alone do not tell the whole story. The impacts are systemic and compounding, triggering cascading failures across ecological, climatic, and infrastructural systems.

Marine heatwaves-those unusually hot periods in the ocean-aren't rare anymore. They're becoming the new normal (Oliver et al., 2018). In 2023 alone, over half of the ocean's surface was experiencing these extreme conditions (Blunden et al., 2024). And it's not just a problem for fish or coral reefs-this has real consequences for all of us. When the ocean heats up like this, it disrupts currents, lowers oxygen levels, and destroys coral ecosystems that support entire marine food chains (Oliver et al., 2018). That means fisheries can collapse, seafood becomes scarcer and more expensive, and coastal communities that rely on the ocean for their livelihoods are thrown into crisis. A warming ocean also loses its ability to absorb carbon dioxide and regulate the planet's temperature-one of the key functions that keeps Earth livable (Wang et al., 2025). As that balance breaks down, the planet heats up faster, weather patterns shift even more wildly, and the impacts ripple back on land in the form of stronger storms, altered rainfall, droughts, and floods (Smale et al., 2019). These events compound food shortages, drive up the cost of living, and increase the risk of displacement and conflict (IPCC, 2023).

So, what's happening out in the middle of the ocean isn't actually far away at all. It's showing up in grocery store prices, in insurance premiums, in the resilience (or failure) of food and water supplies, and in whether people can safely stay where they live. The ocean is changing fast-and when it struggles, the rest of us feel it in ways we're only beginning to understand (NOAA, 2023).

At the same time, the physical landscape of the planet is changing in ways that directly affect *where* people can live. Coastal areas are eroding due to rising sea levels and stronger storms, leading to land loss and increased flood risk. Your lovely beachfront property is likely to turn a little more beach *in* property than you might like. In many regions, farmland is drying out as temperatures rise and rainfall becomes less

[1] Reproduced from Foster & Rahmstorf (2025), Table 2.1. Some modifications were made. https://doi.org/10.21203/rs.3.rs-6079807/v1, licensed under the terms of the Creative Commons Attribution 4.0 International License (http://creativecommons.org/licenses/by/4.0/).

reliable (IPCC, 2022). Droughts are lasting longer, and soil is degrading, making it harder to grow food. And food is kind of important for us humans last time I checked.

These changes are shifting the boundaries of habitability. Places that once supported stable communities are becoming less viable due to these repeating combinations of heat stress, water scarcity, and declining agricultural productivity (IPCC, 2022). As these pressures grow, more people will be forced to relocate, and essential systems like food supply and water access will come under increasing strain (ibid.). A perfect cocktail for migration crises and conflict.

Weather extremes as described earlier are no longer isolated or seasonal. Heatwaves, megastorms, wildfires, droughts, and floods are colliding more often, feeding into one another in destructive cycles (IPCC, 2023). A storm that follows a drought, or a flood that hits already-scorched land, causes far more damage than either would alone. This is the new reality: climate volatility as a defining feature of the Earth system, with disruption happening faster than many systems can absorb or recover from (IPCC, 2023). If they can at all.

Even the critical circulation systems that regulate Earth's climate are beginning to show signs of instability. One of the most vital is the Atlantic Meridional Overturning Circulation (AMOC)-a large system of ocean currents that moves warm, salty water from the tropics northward at the surface, and returns cooler water southward at depth. This circulation plays a key role in regulating climate, especially across Europe, North America, and West Africa, by helping to balance heat and moisture across the Atlantic Ocean. Evidence now shows that the AMOC may be weakening, and could slow down significantly or even collapse entirely if warming and ice melt continue at current rates. Some studies suggest this could happen as early as the mid-twenty-first century, making it a plausible risk within the lifetime of today's children (Smolders et al., 2024).

If the AMOC weakens further, the consequences could be severe and far-reaching. In Northern and Western Europe, winters could become much colder and stormier, despite global warming. West Africa and parts of South America could experience significant droughts, disrupting agriculture and water supplies. Monsoon systems in South Asia and parts of Africa could shift or weaken, putting food production for hundreds of millions of people at risk. In broader terms, an AMOC collapse would not just change regional climates—it would destabilize weather systems across the globe, making them more erratic and harder to predict. This would further challenge efforts to plan for agriculture, manage water supplies, and protect infrastructure from temperature changes. How do we plan for both the hottest and the coldest of climates? The weakening of this system is a stark reminder that the Earth's climate is not just gradually warming-it is moving toward non-linear and potentially irreversible shifts that would have direct and widespread impacts on people's lives across continents. It may not happen "The Day After Tomorrow" or in the way it does in the movie, but it's looking more likely to happen at some point.

With or without AMOC collapse, the energy systems we use to heat and cool ourselves are creaking under the strain. As temperatures rise, so does demand for air conditioning, even as the ability to generate and distribute electricity declines. Heatwaves cause surges in demand, and the potential for breakdown of electricity grids just when people need energy the most (IPCC, 2022). Imagine in the hottest places in the world, air conditioning failing as the hottest temperatures rip through towns and cities. We don't need to, it's already happening. The consequences can be large numbers of heat-related deaths, which people might have thought they could avoid with this technology. But the technology is still electricity dependent. The places we live-the buildings, roads, and houses we live in-were not built for this. Much of it was designed for a stable past, not the volatile present or the even more unstable future (Davis, 2024). Infrastructure across both cities and rural areas risks failing more often and more dramatically under the weight of how extreme the climate is actually becoming.

And as the climate shifts, so do the patterns of disease. Warmer temperatures, changing rainfall, and longer growing seasons are allowing insect- and water-borne diseases to spread into new regions and remain active for longer parts of the year. Mosquitoes have recently been identified in Iceland for the first time in its history due to the warming temperatures! At the same time, worsening air and water quality-driven by a combination of pollution, heat, and drought-are contributing to a rise in respiratory conditions, cardiovascular stress, and gastrointestinal infections, particularly in areas with already limited access to healthcare (IPCC, 2022).

These health impacts are occurring alongside rising threats to food systems. Health services must respond to emerging disease risks, while farmers and food supply chains struggle to adapt to more hostile growing conditions. The result is a feedback loop of vulnerability: reduced food production can lead to higher prices and nutritional stress, which in turn weakens public health-especially in low-income or climate-vulnerable regions. The impacts of a changing climate are not isolated-they are interconnected, with wide-reaching consequences for both human and environmental wellbeing (Heinzel, 2023).

Even before the direct physical impacts of climate change occur, the psychological burden is already growing. A growing body of research shows that climate-related mental health issues-including anxiety, depression, grief, and trauma-are on the rise. This phenomenon, sometimes referred to as eco-anxiety, climate distress, or solastalgia (the emotional pain caused by environmental change), is being reported globally, particularly among young people, farmers, Indigenous communities, and those living in disaster-prone areas (Ciancioni et al., 2020). A 2021 global study published in *The Lancet* found that 59% of young people surveyed across 10 countries reported feeling "very" or "extremely" worried about climate change, and 45% said these concerns negatively affect their daily life and functioning (Hickman et al., 2021). The constant exposure to reports of wildfires, floods, species loss, and failed harvests contributes to a growing sense of fear, helplessness, and uncertainty about the future. This psychological stress is not limited to those directly experiencing disasters. It is also affecting individuals who feel powerless in the face of global environmental breakdown. Mental health professionals are reporting increases in patients presenting with climate-related worry and despair, even in high-income countries that are not yet experiencing the worst impacts (Ingraham, 2024).As climate impacts accelerate, the need to recognize and address the mental health dimension of the crisis becomes increasingly urgent-both as a public health issue and as a component of broader climate adaptation strategies (American Psychological Association, 2021).

All of this is somewhat inevitably beginning to force people to move. As conditions deteriorate, migration-both sudden and slow-is accelerating. By 2050, between 44 and 143 million people could be on the move within their own countries due to crop failures, water scarcity, and rising seas (Kanta et al., 2018). Some models estimate that by 2070, as many as 3.5 billion people could find themselves in regions too hot for sustained human life. That scale of movement would be unprecedented-and it would place immense pressure on already strained governance systems and the supplies needed for survival (Kulp & Strauss, 2019; Fleming, 2020).

Conclusion

If you're still reading, thank you-and well done. That was heavy, and it was meant to be. Facing the reality of what's unfolding isn't easy, and writing about it wasn't either. But this is where we are.

The important thing now is how we respond. In the chapters ahead, we'll shift focus toward what we can actually do-not to fix everything, but to adapt, prepare, and protect what matters most. We may not have control over the bigger systems driving this crisis, but we do have choices in how we respond, support others, and build resilience together.

There is strength in facing the truth, and there is power in being prepared. Let's begin there.

Reflection and Discussion Questions

1. What early signs of climate stress (e.g., food price increases, weather extremes, mental health impacts) can you identify in your own region or profession? How could you begin tracking or mapping these changes locally?
2. Which community-based strategies (e.g., mutual aid groups, food-growing networks, resource-sharing, and climate support circles) could you join or help initiate in response to the vulnerabilities outlined in this chapter?
3. How might you start a conversation—at work, in your family, or in your community—about planning for a future that includes disruption and instability? What tools (e.g., workshops, climate cafés, scenario planning) could support that dialogue?
4. If you work in a profession like social work, education, or health care, what is *one concrete adaptation* you could propose or begin implementing today to build local climate resilience? Think small-scale but high-impact (e.g., community gardens,

emergency planning, mental health peer support, and training in trauma-informed care).

5. What personal or household-level steps could you take now to increase your resilience (e.g., food growing, water storage, skills learning, network building)? How can you do this in a way that includes and supports others, especially those more vulnerable?

References

American Psychological Association & ecoAmerica. (2021). *Mental health and our changing climate: Impacts, inequities, and pathways to resilience* [Report]. https://www.apa.org/news/press/releases/2021/09/mental-health-climate.pdf

Blunden, J., Arndt, D. S., & Hartfield, G. (Eds.) (2024). State of the climate in 2023. *Bulletin of the American Meteorological Society, 105*(8).

Cianconi, P., Betrò, S., & Janiri, L. (2020). The impact of climate change on mental health: A systematic descriptive review. *Frontiers in Psychiatry, 11*, 74. https://doi.org/10.3389/fpsyt.2020.00074

Davis, N. (2024, March 7). Weatherwatch: Pothole problem to deepen amid climate crisis. *The Guardian.* https://www.theguardian.com/world/2024/mar/07/weatherwatch-pothole-problem-deepen-climate-crisis

Fleming, S. (2020, May). Third of world's population may live in Sahara-like heat zones by 2070. *World Economic Forum.*

Foster, G., & Rahmstorf, S. (2025). *Global warming has accelerated significantly* [Preprint]. Research Square. https://doi.org/10.21203/rs.3.rs-6079807/v1

GreenFacts. (n.d.). *Impacts of a 4 °C global warming.* In *Impacts of global warming.* Retrieved June 28, 2025, from GreenFacts website.

Hansen, J. E., Sato, M., Simons, L., Nazarenko, L. S., Sangha, I., von Schuckmann, K., et al. (2023). Global warming in the pipeline. *Oxford Open Climate Change.*

Heinzel, V. (2023, February 9). *When plants get sick, we do too.* Axios. Retrieved from https://www.axios.com/2023/02/09/plant-health-food-insecurity

Hickman, C., Marks, E., Pihkala, P., Clayton, S., Lewandowski, E., Mayall, E., Wray, B., Mellor, C., & van Susteren, L. (2021). Climate anxiety in children and young people and their beliefs about government responses to climate change: A global survey. *Lancet Planetary Health, 5*(12), 863–873. https://doi.org/10.1016/S2542-5196(21)00278-3

Ingraham, C. (2024, January 4). *Minnesota mental health professionals say climate concerns driving patients to depression.* Minnesota Reformer. Retrieved from https://minnesotareformer.com/2024/01/04/mn-mental-health-professionals-say-climate-concerns-driving-many-patients-to-severe-depression/

Intergovernmental Panel on Climate Change. (2018). *Global warming of 1.5 °C: An IPCC Special Report on the impacts of global warming of 1.5 °C above pre-industrial levels and related global greenhouse gas emission pathways.* IPCC.

Intergovernmental Panel on Climate Change. (2021). *Climate change 2021: The physical science basis. Contribution of Working Group I to the sixth assessment report of the intergovernmental panel on climate change* (V. Masson-Delmotte et al., Eds.). Cambridge University Press. https://www.ipcc.ch/report/ar6/wg1/

Intergovernmental Panel on Climate Change. (2022). *Summary for policymakers* [from *Climate Change 2022: Impacts, Adaptation and Vulnerability*, Working Group II contribution]. Cambridge University Press. https://doi.org/10.1017/9781009325844.001

Intergovernmental Panel on Climate Change. (2023). *Climate change 2023: Synthesis report. Contribution of Working Groups I, II and III to the Sixth assessment report of the intergovernmental panel on climate change.* https://www.ipcc.ch/report/ar6/syr/

de Kanta, A., Sherbinin, B., Jones, S., Adamo, D., Maleki, A. A., & Casals Fernández, T. (2018). Groundswell: Preparing for internal climate migration. *World Bank.* https://doi.org/10.1596/978-1-4648-1328-3

Kaufman, D., McKay, N., Routson, C., Erb, M., Dätwyler, C., et al. (2022). Holocene global mean surface temperature, a multi-method reconstruction approach. *Scientific Data, 7*, 201.

Kulp, S. A., & Strauss, B. H. (2019). New elevation data triple estimates of global vulnerability to sea-level rise and coastal flooding. *Nature Communications, 10*, 4844.

Lenton, T. M., & Williams, H. T. P. (2013). On the origin of planetary-scale tipping points. *Trends in Ecology & Evolution, 28*(7), 380–382. https://doi.org/10.1016/j.tree.2013.05.001

Lynas, M. (2007). *Six degrees: Our future on a hotter planet*. Fourth Estate.

NASA. (2020). *Climate change: How do we know?* https://climate.nasa.gov/evidence/

NOAA. (2023). *The ocean is our climate: Why the ocean matters in a warming world*. National Oceanic and Atmospheric Administration. https://www.noaa.gov/ocean-climate

Oliver, E. C. J., Donat, M. G., Burrows, M. T., Moore, P. J., Smale, D. A., et al. (2018). Longer and more frequent marine heatwaves over the past century. *Nature Communications, 9*, 1324. https://doi.org/10.1038/s41467-018-03732-9

Potts, R. (1998). Environmental hypotheses of hominin evolution. *American Journal of Physical Anthropology, 107*(S27), 93–136.

Richerson, P. J., & Boyd, R. (2005). Was agriculture impossible during the Pleistocene but mandatory during the Holocene? *American Antiquity, 70*(4), 523–554.

Sherwood, S. C., & Huber, M. (2010). An adaptability limit to climate change due to heat stress. *Proceedings of the National Academy of Sciences of the United States of America, 107*(21), 9552–9555. https://doi.org/10.1073/pnas.0913352107

Simons, L. (2025, August). *Table assessing linear projections forward based on the new data from Grant Foster & Stefan Rahmstorf* [Table]. LinkedIn. https://www.linkedin.com/posts/leon-simons-b715989_the-most-important-insight-from-these-adjusted-activity-7360616522668462080GOkN?utm_source=share&utm_medium=member_desktop&rcm=ACoAAC07yEoBBx5SOktkQUIV34Ivcnpa_eL38Mg

Smale, D. A., Wernberg, T., Oliver, E. C. J., Thomsen, M., Harvey, B. P., Straub, S. C., & Burrows, M. T. (2019). Marine heatwaves threaten global biodiversity and the provision of ecosystem services. *Nature Climate Change, 9*, 306–312. https://doi.org/10.1038/s41558-019-0412-1

Smolders, E. J. V., van Westen, R. M., & Dijkstra, H. A. (2024). *Probability estimates of a 21st century AMOC collapse.* Advance online publication. https://doi.org/10.5194/gmd-2024-123

Wang, Z., Li, X., Song, J., Zhong, G., Liu, S., He, Z., & Duan, L. (2025). A new indicator can assess absorption capacity for carbon dioxide and ocean acidification. *Communications Earth & Environment, 6*(1), 80. https://doi.org/10.1038/s43247-025-02380-4

World Meteorological Organization. (2024). *State of the global climate 2023.* Geneva, Switzerland. Retrieved from https://public.wmo.int/en/resources/library/state-of-global-climate

Chapter 3
Reimagining Prevention: Integrating Climate Adaptation into Everyday Social Work Practice

Introduction

Whether they realize it or not, every social worker is already working within a changing climate. What we see in our day-to-day work include vulnerabilities created by food insecurity, driven by high food prices, exacerbated by harvest failures. We work with older adults and children experiencing the effects of heat stress in summer, and notice how housing is not adapted to the unprecedented changes in temperature and extreme weather. The practice, methods, and interventions, however, have not caught up with the changes. We have not fully attempted to do more than acknowledge the increased vulnerabilities climate change brings, but what is needed is to embed climate adaptation as a form of anticipatory social work.

What's missing from most preventive social work is a deliberate focus on climate adaptation—helping people and communities prepare for, live with, and recover from environmental change. Many of our current frameworks still assume that stability will return after each crisis, that there will always be time and resources to rebuild. Even those workers operating in warzones and disaster affected areas have the possibility of tapping into external stable regions for international aid for example. Yet even the most conservative climate projections tell a different story. The IPCC (2023) warns of continued warming, more frequent extreme weather, and the loss of the natural systems that once buffered us from disaster. Ripple et al. (2024) and Hansen et al. (2025) describe this as an accelerating decline- a trend that will shape every aspect of human wellbeing. No one may be able to come and help when we need it.

For social work, this means operating in conditions that may remain unstable for the foreseeable future, perhaps forever. Services designed for steady systems will face repeated shocks. Here I acknowledge that there are many places in the world where the systems in which people practice social work are already unsteady, in which they are already having to adapt due to conflict or poverty aside from climate impacts. The profession's capacity to protect and support vulnerable groups as we move forward could be stretched to breaking point (Serdeczny et al., 2024). As Dominelli (2014)

S. Rost, *Climate Resilience Guide for Social Workers*,
SpringerBriefs in Social Work, https://doi.org/10.1007/978-3-032-24604-2_3

has cautioned, unless we act before crises occur, we risk being trapped in a constant cycle of emergency response.

The law concerning social work already places a focus on early intervention or preventive practices in many countries.

Examples of preventive social work embedded in law:

Scotland

- Getting it right for every child (GIRFEC) is the national framework that embeds *prevention and early intervention* across children's services (education, health, social work) to uphold wellbeing and deliver support "when needed," not only at crisis. (Scottish Government, 2023)

Norway

- The Public Health Act (2011) gives municipalities a legal mandate for *long-term, systematic public health and prevention*, requiring "health in all policies," local public-health overviews, and integration of social determinants (e.g., housing, education) into municipal plans. This shifts prevention into routine local governance (Norwegian ministry of health & care services, 2011).

New Zealand (Aotearoa)

- The Oranga Tamariki Act 1989 (children's law) centers wellbeing and establishes principles that drive *early intervention* for care and protection-operationalized through cross-agency early-intervention programs and guidance (New Zealand Parliament, 1989).

Why this is important is that intervening in an anticipatory way is something that is not only within the social work remit, but existing laws can be used to create a focus on climate adaptation as a form of early intervention.

Preventive social work has several key foci or intent in terms of what it hopes to achieve as outcomes. These include preventing social problems such as child harm, family breakdown, crime and mental health issues among others by addressing the root causes early, or addressing them at an early enough stage prior to the negative effect (Tollan et al., 2023). This involves equipping individuals and families with skills, social supports, and programs designed to reduce dependency on social services (Ibid). It includes home visits, youth outreach, parenting programs, family counseling, school-based support, and mental health and substance use prevention, all aimed at promoting stability as well as the use of time limited focused programs (Winter et al., 2021). Community-based approaches foster resilience through social cohesion, while housing security efforts focus on preventing eviction and homelessness (Fitzpatrik et al., 2021). At a structural level, social workers advocate for policy change to address systemic issues like poverty and inequality (Chapin & Lewis, 2023). Central to effective prevention is collaboration across sectors such as health, education, and social services, with particular attention to at-risk populations (Moran et al., 2007).

We must acknowledge that due to the likely escalation in need brought about through climate impacts, demand for support and services may rise substantially. The workload, already heavy, could push workers out of the profession due to burnout or emotional exhaustion, just as they are needed the most. Helping embed climate adaptation into casework now could ease the pressure by anticipating the potential problems before they arise.

The following are examples of how practice areas can include climate adaptation into their practice.

Practice Areas

Practice Area 1: Early Intervention

Early intervention is about identifying risks before they escalate, and climate change adds a new layer to what "risk" means. Within early intervention, where the goal is to anticipate and mitigate risks before they become entrenched, climate adaptation could readily become a standard part of risk assessments, case planning, and child protection protocols. The IPCC (2014) specifically advises that in order to adapt, we need information on who and what areas are at risk to climate impacts. This necessitates that social workers could begin by identifying geographic and social vulnerabilities, such as exposure to flooding, heatwaves, drought, or food insecurity, and flagging communities most likely to be affected. While concepts such as cloudburst mapping already exist in places such as Gothenburg, Sweden (SMHI, 2021), they do not take into account social vulnerabilities, neglecting potential impacts brought about through poverty, age, disability, and so on.

Elements of this have already begun in health services, with additional checks on vulnerable people during heat waves (Curtis et al., 2017). During home visits or outreach work, social workers could share information on practical steps families can take to reduce risks—such as preparing for power outages, using sandbags to protect homes, growing food locally, or staying cool during extreme heat. For families without the means to take these steps, early intervention could include financial assistance, coordination with municipal services, or linking clients with community emergency response networks to ensure no one is isolated or unsupported during climate-related disasters. These approaches reflect international bodies' advice that local preparedness is vital (United Nations, n.d.).

By putting in preventative measures that take into account the risks climate pose to families, early intervention can not only fulfill its purpose of prevention of harm but extend this to include impacts brought about by climate disruption.

Practice insight: Social workers may notice familiar cases returning more often, clients showing signs of chronic stress, or communities relying more heavily on informal mutual aid. Recognizing these as climate-related symptoms helps shift practice from crisis response to long-term adaptation.

Practice Area 2: Family and Child Support Services

Family-focused practice increasingly needs to account for the psychological and social strain created by environmental change. Uncertainty about the future, rising costs, and displacement can all increase stress and conflict within families. Evidence links extreme heat and other climate stressors to declines in mental well-health and higher suicide rates (Li et al., 2022; Mason et al., 2018). Preventive responses might include creating resilience through knowledge sharing and emotional support. Universalising the emotions felt by those experiencing impacts of climate and setting up groups for people to talk about their fears could mean people would feel less alone, and therefore reducing vulnerability. And this is important to emphasize—having anxiety as a response to climate issues is a rational response not a pathological one. This is also a key aspect of how climate adaptation and preventive social work can be linked—aim to increase protective factors and reduce risk factors, but to risks specific to climate. For families and children, risks could be housing, financial, emotional as well as environmental. However, there may also be aspects not previously considered as specific vulnerabilities, such as lack of specific skills or knowledge such as those which may help in adaptation or survival. This type of knowledge could be beyond our traditional experience base and challenge us to expand into areas of preparedness. It offers, however, the opportunity to not only share knowledge and skills that may benefit families, but encourage strengths-based approaches to assess more openly what might this person be able to do quite well in a crisis? How might their life experiences contribute to their ability to survive? This could be incorporated into higher education for social workers but also ongoing for those already established in social work settings.

Practice insight: Families may be experiencing risks that are often overlooked or not formally measured. By recognizing how climate change affects people's daily lives, social workers can understand clients more holistically and ensure their practice is truly person-centered.

Practice Area 3: Community-Based Practice

In community-based practice, social cohesion is at risk due to climate changes (Danish Institute for International Studies, n.d.) which reduce effectiveness of community-based preventive social work. However, by promoting climate adaptations within communities, it offers the opportunity to create a sense of "active hope" (Lawrance et al., 2022), and increase social cohesiveness (Tamasiga et. al., 2024). Using climate adaptation in an anticipatory way with community-based practice could function by creating relationships and networks between individuals and groups. This focuses on the strength that the collective can have over that of the individual. Social capital is something that can be forged after disasters, but it could also be developed or the foundations put in place in advance, so that networks can fall into place through preplanned arrangements. This can include use of human mesh networks—human groups of contacts that can be activated in an emergency (Rost, in press). Telephone groups, community hubs, and meeting points can be used as central points of information. Existing social work premises such as youth centers, food banks, or offices can be used as bases for cooling centers in heatwaves, warming centers in winter storms or places that can offer skills training in different aspects of climate adaptation. This can include skills in how to conserve water, grow food or provide services to each other as mutual aid. These local actions demonstrate how social work can nurture the resilience of both people and place (Danish Institute for International Studies, n.d.).

Practice insight: Communities may offer the best support to the most vulnerable, as your neighbors and those in the immediate vicinity are likely to be the ones to help you in an emergency or disaster. Cultivating those relationships asking your clients who they can reach out to may help promote their resilience.

Practice Area 4: Education and School-Based Practice

Schools represent key spaces for building climate resilience, not only for children but for entire communities. Because they reach young people across diverse social and economic backgrounds, they offer a practical entry point for social workers to integrate resilience education into everyday life (Williams & McEwen, 2021). By collaborating with educators, social workers can help design activities that strengthen adaptive capacity such as teaching food-growing skills, water conservation methods, community cooperation, and creative responses to environmental disruption. Knowledge becomes a central tool of resilience: when children understand how to adapt, they are better equipped to safeguard both their own wellbeing and that of their families and neighborhoods. These experiences cultivate both immediate, practical understanding and intergenerational learning that can be passed forward to their families and their own children (Williams & McEwen, 2021).

In addition, school-based social workers can use social and emotional learning programs to explore climate-related emotions and coping strategies, providing psychological grounding alongside practical preparation. Young people are already acutely aware of the climate crisis and often express anxiety about their future (Whitlock, 2023). Supporting them through open dialogue and action-oriented education helps transform fear into agency, rather than simply pretending to them that the problem doesn't exist. As the IPCC (2023) note, the challenges of climate change will increasingly shape children's lived realities. Equipping them with ways to act-individually and collectively-can reduce helplessness (Geiger et al., 2021) and foster a proactive resilience.

Practice insight: "Engaging with green spaces supports positive mental health in youth by building their resilience, sense of autonomy and positive outlook" (Lam et al., 2019, p. 1). These can be physical green spaces such as community gardens or places where discussion on climate issues is offered.

Practice Area 5: Mental Health and Substance Use Support

From a social work resilience perspective, mental health and substance use services must be strengthened to withstand the growing pressures of climate change. As environmental instability intensifies, social workers will increasingly encounter clients whose anxiety, trauma, and substance use are linked to climate-related stressors such as displacement, economic insecurity, and heat-related distress. Research shows that substance misuse tends to rise under these conditions, particularly among young people, alongside elevated rates of mental health challenges (Vergunst et al., 2022).

Building resilience within these systems requires more than physical adaptation—it demands service continuity, flexibility, and trauma-informed preparedness. Social workers can play a key role in advocating for climate-resilient infrastructures, including backup energy systems, flood protection, green cooling spaces, and access to emergency supplies. These measures are essential to ensure that vulnerable clients can continue receiving care during and after climate events.

Climate change also disrupts the social and material foundations of treatment. Interrupted supply chains of both illegal and legal drugs can lead to contaminated or adulterated substances and shortages of essential medications such as methadone, compounding risk for people with substance use disorders (Canadian Centre on Substance Use & Addiction, 2024). Health centers and offices may face infrastructure breakdowns, cutting off access to essential services. Social workers can strengthen system resilience by helping to develop adaptive service models-for example, mobile mental health units, local reserves of essential medicine in parallel with health services, and create digital or telehealth networks that maintain contact when physical infrastructure fails.

Resilient social work practice means anticipating disruptions, maintaining relational continuity with clients, and adapting service delivery to ensure no one is left without support in moments of crisis.

Practice insight:

Preparation for shortages of medicines may not appear to be part of the job description for social workers, but understanding the consequences to vulnerable populations that may be affected can help you be more anticipatory in your assessments of potential needs.

Practice Area 6: Housing and Shelter

Extreme weather events, such as floods, storms, and sea-level rise, are already disrupting housing security and displacing families, intensifying strain on social systems (Ajibade et al., 2021). For social workers, this requires a shift from crisis response to preventive climate preparedness, ensuring that safe and accessible shelters and housing are available before disaster strikes. This not only acts as a preventive measure for the future but also offers support for those already experiencing homelessness or housing precarity (Amorim-Maia et al., 2023). This can take the form of a stronger advocacy for housing support and closer relationship with these services in order to try to help those affected but can also be identifying hazards such as mold or lack of ventilation or heating and cooling for those in precarious housing.

Preventive adaptation can also include the creation of climate-responsive spaces, such as cooling centers during heatwaves, which can be lifesaving for older adults, people with chronic illness, or those living outdoors. Similarly, in winter this can be used for freezing temperatures that may put lives at risk. Beyond the physical, social workers must recognize and respond to the emotional and relational dimensions of displacement. Climate-induced migration, whether within a country or across borders, brings profound loss not just of homes, but of social networks and a sense of belonging. Providing trauma-informed and culturally sensitive support helps individuals rebuild a sense of stability and belonging amid environmental and social upheaval (Ekoh et al., 2023). Helping people make connections to groups in the area can promote social capital and cohesion even amid crises.

Practice insight: You don't need to be a town planner to have an understanding of which areas are most likely to flood or experience greatest heat risks during heatwaves. Mapping these even to some degree can help identify vulnerable groups *before* a disaster occurs.

Practice Area 7: Advocacy and Policy Application

Climate-resilient social work extends beyond direct practice to include systems-level advocacy, where influencing policy becomes a central form of adaptation. Social workers are uniquely positioned to work alongside local governments, ensuring that climate strategies address the realities of those most affected by environmental and social inequities. Acting preventively at this level means advocating for adaptation measures that prioritize justice and inclusion-for instance, supporting investment in green areas, community gardens, and food forests, and implementing cool spaces in built up areas (Curtis et al., 2017).

By collaborating with organizations focused on food security and sustainable agriculture, social workers can help strengthen localized and resilient food systems, reducing the vulnerability of people in their area to food insecurity (Horn et al., 2022).

Practice insight: By having an anticipatory stance in your practice, you are uniquely positioned to advocate for those who may be vulnerable to climate impacts. Plan for what might happen.

Case Studies

Case Study 1: Heatwaves and Housing Insecurity

A social worker in a midsized city is supporting several families in substandard housing. A prolonged summer heatwave is expected, with temperatures exceeding 40 °C. One family includes an elderly woman with health issues, a single parent with young children, and a recently arrived refugee family living in an overcrowded apartment without air conditioning.

Questions:

- What immediate and long-term adaptation strategies could the social worker implement or advocate for?
- How might this social worker collaborate with local services or community groups?
- What are the barriers to action, and how could they be overcome?

Case Study 2: Drought and Substance Use in a Rural Community

A rural town is experiencing long-term drought. Incomes have fallen, mental health issues are rising, and there's been an increase in substance use among youth. Local clinics are understaffed and struggling with medication shortages.

Questions:

- How could preventive social work approaches respond to this intersection of environmental, social, and health challenges?
- What role might community-based practice and advocacy play here?
- What systemic or policy-level interventions could be considered?

Case Study 3: School Resilience and Eco-Anxiety

A school-based social worker notices increasing anxiety among students about climate change. Some students express fear of the future, while others are disengaged and pessimistic. The school has no current framework for addressing environmental concerns or adaptation education.

Questions:

- How can the social worker support students' mental wellbeing while fostering resilience and agency?
- What partnerships or curricular adaptations might support long-term change?
- How would you balance climate realism with emotional care?

References

Ajibade, I. J., & Siders, A. R. (Eds.). (2021). *Global views on climate relocation and social justice: navigating retreat* (1st ed.). Routledge. https://doi.org/10.4324/9781003141457

Amorim-Maia, A. T., Anguelovski, I., Connolly, J., & Chu, E. (2023). Seeking refuge? The potential of urban climate shelters to address intersecting vulnerabilities. *Landscape and Urban Planning, 238*, Article 104836. https://doi.org/10.1016/j.landurbplan.2023.104836

Canadian Centre on Substance Use and Addiction. (2024). *Substance use and climate change: A comprehensive review.* https://www.ccsa.ca/sites/default/files/2024-02/Substance-Use-and-Climate-Change-en.pdf

Chapin, R. K., & Lewis, M. (2023). *Social policy for effective practice: a strengths approach* (6th ed.). Routledge. https://doi.org/10.4324/9781003273479

Curtis, S., Fair, A., Wistow, J., et al. (2017). Impact of extreme weather events and climate change for health and social care systems. *Environmental Health, 16*(Suppl 1), 128. https://doi.org/10.1186/s12940-017-0324-3

Danish Institute for International Studies (DIIS). (n.d.). *Climate-related losses and damages to social cohesion are overlooked.* https://www.diis.dk/en/research/climate-related-losses-and-damages-to-social-cohesion-are-overlooked

Dominelli, L. (2014). Promoting environmental justice through green social work practice. *International Social Work, 57*(4), 338–345.

Ekoh, P. C., Iwuagwu, A. O., George, E. O., & Walsh, C. A. (2023). Forced migration-induced diminished social networks and support, and its impact on the emotional wellbeing of older refugees in Western countries: A scoping review. *Archives of Gerontology and Geriatrics, 105*, Article 104839. https://doi.org/10.1016/j.archger.2022.104839

Fitzpatrick, S., Mackie, P., & Wood, J. (2021). Advancing a five-stage typology of homelessness prevention. *International Journal on Homelessness, 1*(1), 79–97.

Geiger, Swim, J. K., Gasper, K., Fraser, J., & Flinner, K. (2021). How do I feel when I think about taking action? Hope and boredom, not anxiety and helplessness, predict intentions to take climate action. *Journal of Environmental Psychology*, *76*, 101649. https://doi.org/10.1016/j.jenvp.2021.101649

Hansen, J. E., Sato, M., & Ruedy, R. (2025). Global warming has accelerated: Are the United Nations and the media reporting the real climate story? *Environment: Science and Policy for Sustainable Development*, *67*(1), 4–15. https://doi.org/10.1080/00139157.2025.2434494

Horn, B., Ferreira, C., & Kalantari, Z. (2022). Links between food trade, climate change and food security in developed countries: A case study of Sweden. *Ambio, 51*(4), 943–954. https://doi.org/10.1007/s13280-021-01623-w

Intergovernmental Panel on Climate Change. (2014). *Climate Change 2014: Impacts, Adaptation, and Vulnerability. Part A: Global and Sectoral Aspects. Contribution of Working Group II to the Fifth Assessment Report of the Intergovernmental Panel on Climate Change* (C. B. Field et al., Eds.). Cambridge University Press. https://www.ipcc.ch/site/assets/uploads/2018/02/WGIIAR5-PartA_FINAL.pdf

Intergovernmental Panel on Climate Change (IPCC). (2023). *Climate change 2023: Synthesis report of the IPCC sixth assessment report* [Longer report]. https://www.ipcc.ch/report/ar6/syr/downloads/report/IPCC_AR6_SYR_LongerReport.pdf

Lam, V., Romses, K., & Renwick, K. (2019). Exploring the relationship between school gardens, food literacy and mental well-being in youths using Photovoice. *Nutrients, 11*(6), Article 1354. https://doi.org/10.3390/nu11061354

Lawrance, E. L., Thompson, R., Newberry Le Vay, J., Page, L., & Jennings, N. (2022). The Impact of Climate Change on Mental Health and Emotional Wellbeing: A Narrative Review of Current Evidence, and its Implications. *International Review of Psychiatry, 34*(5), 443–498. https://doi.org/10.1080/09540261.2022.2128725

Li, C., Lawrance, E. L., Morgan, G., Brown, R., Greaves, N., Krzanowski, J., & Belkin, G. (2022). The role of mental health professionals in the climate crisis: An urgent call to action. *International Review of Psychiatry, 34*(5), 563–570. https://doi.org/10.1080/09540261.2022.2097005

Mason, L. R., Erwin, J., Brown, A., Ellis, K. N., & Hathaway, J. M. (2018). Health IMPACTS OF EXTREME WEATHER EVENTS: EXPLORING PROTECTIVE FACTORS WITH A CAPITALS FRAMEWORK. *Journal of Evidence-Informed Social Work, 15*(5), 579–593. https://doi.org/10.1080/23761407.2018.1502115

Moran, P., Jacobs, C., Bunn, A., & Bifulco, A. (2007). Multi-agency working: Implications for an early-intervention social work team. *Child & Family Social Work, 12*(2), 143–151. https://doi.org/10.1111/j.1365-2206.2006.00452.x

New Zealand Parliament. (1989). *Oranga Tamariki Act 1989* (formerly *Children, Young Persons, and Their Families Act 1989*). Retrieved from https://www.legislation.govt.nz/act/public/1989/0024/latest/whole.html

Norwegian Ministry of Health and Care Services. (2011b). *Health and Care Services Act (LOV-2011–06–24–30): Act relating to municipal health and care services.* Government of Norway. https://lovdata.no/dokument/NLE/lov/2011-06-24-30

Ripple, W. J., Wolf, C., Gregg, J. W., Rockström, J., Mann, M. E., Oreskes, N., Lenton, T. M., Rahmstorf, S., Newsome, T. M., Xu, C., Svenning, J.-C., Pereira, C. C., Law, B. E., & Crowther, T. W. (2024). The 2024 state of the climate report: Perilous times on planet Earth. *BioScience, 74*(12), 812–824. https://doi.org/10.1093/biosci/biae087

Rost, S. (in press). *Human mesh networks: Communicating social work research during collapse conditions. Social Work.*

Scottish Government. (2023). *Getting it right for every child (GIRFEC): Policy and practice guidance.* Scottish Government. https://www.gov.scot/policies/girfec

Serdeczny, O., Andrijevic, M., Fyson, C., et al. (2024). Climatic risks to adaptive capacity. *Mitig Adapt Strateg Glob Change, 29*, 10. https://doi.org/10.1007/s11027-023-10103-3

Swedish Civil Contingencies Agency (MSB). (n.d.). *Advice for individuals.* https://www.msb.se/en/advice-for-individuals/

Swedish Meteorological and Hydrological Institute. (2021, October 28). *Översvämning i framtida klimat.* https://www.smhi.se/professionella-tjanster/klimat-och-klimatanpassning/oversvamning-i-framtida-klimat

Tamasiga, P., Mogomotsi, P. K., Onyeaka, H., & Mogomotsi, G. E. J. (2024). Amplifying climate resilience: The impact of social protection, social cohesion, and social capital on public support for climate change action. *Sustainable Environment, 10*(1). https://doi.org/10.1080/27658511.2024.2361568

Tollan, K., Jezrawi, R., Underwood, K., et al. (2023). A Review on Early Intervention Systems. *Current Developmental Disorders Reports, 10*, 147–153. https://doi.org/10.1007/s40474-023-00274-8

United Nations. (n.d.). *Water and climate change.* https://www.un.org/en/climatechange/science/climate-issues/water

Vergunst, F., Berry, H. L., Minor, K., & Chadi, N. (2022). Climate change and substance-use behaviors: a risk-pathways framework. *Perspectives on Psychological Science, 18*(4), 936–954. https://doi.org/10.1177/17456916221132739(Originalworkpublished2023)

Whitlock, J. L. (2023). Climate change anxiety in young people. *Nature Mental Health, 1*(5), 297–298. https://doi.org/10.1038/s44220-023-00059-3researchgate.net+2jedfoundation.org+2

Williams, S., & McEwen, L. (2021). 'Learning for resilience' as the climate changes: Discussing flooding, adaptation and agency with children. *Environmental Education Research, 27*(11), 1638–1659. https://doi.org/10.1080/13504622.2021.1927992

Winter, K., Neeson, L., Sweet, D., Smith, A., Millen, S., & Connolly, P. (2021). Social innovation and social work: A case study of the Early Intervention Support Service. *The British Journal of Social Work, 51*(8), 2872–2891. https://doi.org/10.1093/bjsw/bcaa125

Chapter 4
How the Natural World Fits in to Social Work Climate Resilience

Introduction

To write about climate resilience in social work, without discussion of our place in the natural world would be more than a strange oversight. It would undermine the very idea of resilience to climate change and its impacts. The reason I say this is because to fully understand why climate change is happening, along with the destruction and collapse of eco systems around us, we must first comprehend our place in the world and as part of it. We, like all species on earth, are made from the life "stuff" of the planet on which we live, the same elements, molecular DNA and so in this way we are intimately connected to the place we call home. Humans tend to long for contact with natural environments, a tendency known as biophilia (Wilson, 1984), which reflects our innate bond with the natural world. In terms of the practical aspects of how this knowledge can contribute to resilience as social workers, it offers the prospect that connecting to the earth in more tangible ways may offer something that individualism and disconnection have lost. These include methods of relating to plants, animals, and our surroundings as well as nature-based forms of social work that exist in various forms all over the world.

Social work increasingly recognizes that human wellbeing cannot be separated from the environments that sustain it. Yet despite this awareness, the natural world has rarely been placed at the center of the profession's ethical identity or everyday methods (Noble, 2016). Over recent decades, a number of related approaches such as Green Social Work (Dominelli, 2012), Disaster Social Work (Boetto, 2024), Eco-social work (Närhi, 2004), Ecofeminism (d'Eaubonne, 2022), and Environmental Social Work (Gray et al., 2013) have sought to redress that imbalance. Each recognizes, in its own way, that environmental and social crises are deeply connected and must be addressed together.

However, much of this thinking emerged during a period of relative planetary stability, built on assumptions of recovery, continuity, and external support. These frameworks reflect what might be described as Holocene-era assumptions of stability,

S. Rost, *Climate Resilience Guide for Social Workers*,
SpringerBriefs in Social Work, https://doi.org/10.1007/978-3-032-24604-2_4

a belief that institutions and ecosystems will rebound after disruption. From the standpoint of the Anthropocene, such confidence feels increasingly misplaced. Climate disruption is no longer a series of temporary events but ongoing conditions shaping everyday life (IPCC, 2023; WMO, 2025).

The reliance on Holocene conditions makes many of these theories potentially ill-suited to the scale and persistence of today's crisis. In fairness, no framework could have fully anticipated the unprecedented convergence of environmental and social breakdowns now unfolding. Our human comprehension is shaped by what we know and have experienced previously (Bransford & Johnson, 1972). Dealing with crises as single events can no longer keep pace with compounding hazards, and approaches that depend mainly on ethical orientation or post-impact response may falter when each disaster weakens the very systems meant to respond. Floods follow droughts, fires follow storms, and each will likely erode physical, emotional, and institutional capacities, including the systems on which social work itself relies.

Yet these frameworks still offer crucial insights. They remind us of the deep importance of place, the planet as our mutual home, and interspecies connection. They also signal that social work can and must play a role in supporting those who face escalating climate impacts and the cascading crises that accompany them. Their limitations, rather than being failings, can guide the profession toward more adaptive and anticipatory forms of climate-resilient practice.

Nature-Based Prevention

Nature-based supports have been increasingly reflected in mental health applications that social workers can use (Bratman et al., 2019). Eco-therapeutic practice is an umbrella term encompassing a range of approaches that use interaction with the natural environment to support psychological and emotional wellbeing. Depending on the context, it may be referred to as nature-based care (Nature England, 2017), ecotherapy (Isham et al., 2025), or green social prescribing (NHS England, n.d.). The activities themselves vary from counseling sessions held outdoors, to reflective walks and mindfulness exercises in nature, to more practical forms of interaction such as growing food together, learning survival skills, or observing wildlife (Isham et al., 2025). Obeng et al. (2023) also talk about using animal interventions, surfing therapy, and environmental conversation as therapeutic tools. These were seen to have benefits in terms of using connectedness to nature to improve emotional wellbeing and resilience, particularly for young people who were deemed most at risk (Obeng et al., 2023).

Gardening, horticulture and having access to a garden has been assessed to improve cognitive development and functioning among children, as well as those with Alzheimer's or brain injury (Corley et al., 2024). This is in addition to improvements in sleep and reductions in violent behaviors. Even viewing nature, through a window or pictures of nature scenes has been seen to have an effect on our emotional wellbeing, with looking at pictures of animals and clouds seen as potentially more calming than actually taking a walk in nature (Heinsch, 2012). Just being able to see a small patch of grass through a window was seen as connected to a reduction in familial violence (ibid).

In terms of physical protection from extreme events, nature-based cooling gives social workers low-cost, practical ways to protect people in heat, which is causing high numbers of deaths during heatwaves (WHO, 2024). The basics for cooling the body are shade, airflow, and moisture (WHO, 2008) and can be obtained through use of tree shade (parks, courtyards, street trees) and setting up temporary shade over benches or queue areas. Many countries are now attempting to "green" their cities, in order to keep temperatures down (United Nations Environment Programme, 2021), so making use of these spaces in times of heat to protect vulnerable people is something social workers can do. You can encourage clients to use tree-lined paths for travel when they need to visit social work offices during hot days and schedule visits in the coolest hours where possible. It is important to have this awareness that setting an appointment in a heatwave at the peak heat hours may be unintentionally discriminating against vulnerable people who may not have ways to keep themselves cool while traveling to meet you (Slesinski et al., 2025). It may be useful to set up partnerships with allotments or other green spaces that are usually closed to the public, to open shaded community cooling spots during heat waves.

So how else can we make use of this knowledge to include nature in our social work practice to improve our climate resilience?

We can begin by arguing for or making use of the green spaces and gardens around us, and not be afraid to take social work out of the office. Walking interviews (Carpiano, 2009) could be used to discuss difficult issues while walking in a park, as long as confidentiality is maintained and the client is in agreement. Social prescribing (Leavell et al., 2019) is another form of integrating nature-based interventions through referring to services such as communal gardens and walking clubs, among others, which can also promote social interaction and connection.

In terms of the physical environment of social work offices, these can be augmented fairly easily with images of nature scenes, use screens or digital installations to present images of nature and tranquil scenes that foster a sense of calm and connection. The classic fish tank we often see when visiting the dentist, is there for a reason (Cracknel et al., 2016) and something we can use in social work settings also. Gardening and gardens offer so many preventive mental health and social benefits relevant to social work that we can operationalise their use as part of our practice—set one up in the grounds of the social work premises so that it can be used by clients and staff. Bring plants from home to the office or request donations of plants from garden centers or charities.

Green roofs, green walls, and other planted surfaces offer practical co-benefits for people and buildings. By this I mean growing plants and vegetation on walls and roofs or creating a "wall" of vegetation with fencing. This has been shown to reduce indoor temperatures in summer and improve insulation in winter, lowering energy use and creating cooler, safer spaces during heatwaves (El Menchawy et al., 2025). This may not be something social workers have the time or skills to create themselves, but they can advocate for their inclusion where possible, especially in creating cool refuges for heatwaves that are not electricity dependent.

On a more individual basis, you could encourage creating smaller versions -from windowsill herbs and microgreens to potting plants, this is something you could do alongside clients to reduce stress and anxiety. The creation of micronutrients that people can also have in their home is also beneficial in response to nutritional deficits caused by food insecurity and poverty (FAO et al., 2023). You could also suggest the idea of "green micro breaks" (Aafjes-van Doorn et al., 2020) during difficult meetings such as child protection conferences. Even brief pauses in a nearby green space-whether a garden, courtyard, or small patch of trees- could help lower stress levels and support emotional regulation for both clients and practitioners. Social workers can additionally use short moments of mindfulness with clients as a way to pause and ground before difficult conversations or decisions. The use of mindfulness in social work is evidence based (Maddock et al., 2022) and can be undertaken outdoors or paired with natural elements, noticing sounds of birds, or focusing on the feel of fresh air, to enhance grounding and connection.

Connection through Social capital is one of the most valuable resources available in times of disaster, yet it is rarely built intentionally or maintained when it is most needed (Rapeli, 2018). Relationships, trust, and informal networks can often hold communities together long after formal systems begin to strain or fail. For social workers, this means that strengthening social capital should be seen as part of everyday practice- not only something that happens during a crisis. Nature-based approaches can strengthen this process. Activities such as community gardening, outdoor support groups, and shared green spaces as described previously create natural opportunities for connection. Working together in nature allows people to form relationships that are both emotionally and practically sustaining. Building these networks has always been central to human survival; cooperation, care, and connection to the natural world helped communities endure previous crises, including the ice ages (Kemp, 2025).

Conclusion

All of these interventions are not meant as a replacement for more concrete help of housing, financial support, or risk assessment, but as an integration of the natural world into our practice. If our aim as social workers is to support those we work with to no longer need us, then meaningful prevention strategies that incorporate our intrinsic connection to nature, appear to be an essential element. Nature offers social connection, ways to reduce anxiety, and practical physical support to help us manage crises, as outlined earlier in this chapter.

Practical Exercise

You are an overworked, exhausted social worker with a high caseload and no time to take on more. You care about the environment and believe nature-based supports might help your clients, but you just don't know how to incorporate them. Now I want you to think of just one thing you could do differently tomorrow that would take little effort and feels achievable. What could it be?

References

Aafjes-van Doorn, K., Kamsteeg, C., & Birkhoff, J. (2020). Green micro-breaks: Brief nature exposure interventions in the workplace. *Frontiers in Psychology, 11*, Article 521497.

Boetto, H. (2024). Disaster practice in social work refers to interventions that build disaster resilience and lessen disaster impacts. *International Social Work*. Advance online publication. https://doi.org/10.1177/00208728241269717

Bransford, J. D., & Johnson, M. K. (1972). Contextual prerequisites for understanding: Some investigations of comprehension and recall. *Journal of Verbal Learning and Verbal Behavior, 11*(6), 717–726. https://doi.org/10.1016/S0022-5371(72)80006-9

Bratman, G. N., Anderson, C. B., Berman, M. G., Cochran, B., de Vries, S., Flanders, J., Folke, C., Frumkin, H., Gross, J. J., Hartig, T., Kahn Jr, P. H., Kuo, M., Lawler, J. J., Levin, P. S., Lindahl, T., Meyer-Lindenberg, A., Mitchell, R., …, Daily, G. C. (2019). Nature and mental health: An ecosystem service perspective. *Science Advances, 5*(7), eaax0903. https://doi.org/10.1126/sciadv.aax0903

Carpiano, R. M. (2009). Come take a walk with me: The "go-along" interview as a novel method for studying the implications of place for health and well-being. *Health & Place, 15*(1), 263–272. https://doi.org/10.1016/j.healthplace.2008.05.003

Corley, J., Pattie, A., Deary, I. J., & Cox, S. R. (2024). Gardening and cognitive ageing: Longitudinal findings from the Lothian Birth Cohort of 1921. *Journal of Environmental Psychology, 97*, Article 10236. https://doi.org/10.1016/j.jenvp.2024.102361

Cracknell, D., White, M. P., Pahl, S., Nichols, W. J., & Depledge, M. H. (2016). Marine biota and psychological well-being: A preliminary examination of dose–response effects in an aquarium setting. *Environment and Behavior, 48*(10), 1242–1269. https://doi.org/10.1177/0013916515597512

d'Eaubonne, F. (2022). *Feminism or death* (R. A. Hottell, Trans.). Verso.

Dominelli, L. (2012). *Green social work: From environmental crises to environmental justice.* Polity Press.

El Menchawy, A. S., Elkholy, M. A., & Mohamed, A. F. A. (2025). Applying green roof in an educational building in the hot arid zone to achieve thermal comfort. *Journal of Umm Al-Qura University for Engineering and Architecture*. https://doi.org/10.1007/s43995-025-00209-x

FAO, IFAD, UNICEF, WFP, & WHO. (2023). *The state of food security and nutrition in the world 2023: Urbanization, agrifood systems transformation and healthy diets across the rural–urban continuum.* Food and Agriculture Organization of the United Nations. https://doi.org/10.4060/cc3017en

Gray, M., Coates, J., & Hetherington, T. (Eds.). (2013). *Environmental social work*. Routledge.

Heinsch, M. (2012). Getting down to earth: Finding a place for nature in social work practice. *International Journal of Social Welfare, 21*(3), 309–318. https://doi.org/10.1111/j.1468-2397.2011.00860.x

Isham, A., Jefferies, L., Blackburn, J., Fisher, Z., & Kemp, A. H. (2025). Green healing: Ecotherapy as a transformative model of health and social care. *Current Opinion in Psychology, 62*, Article 102005. https://doi.org/10.1016/j.copsyc.2025.102005

IPCC. (2023). *Climate change 2023: Synthesis report—Summary for policymakers.* IPCC. https://doi.org/10.59327/IPCC/AR6-9789291691647.001

Kemp, L. (2025). *Goliath's curse: The history and future of societal collapse.* Viking.

Leavell, M. A., Leiferman, J. A., Gascon, M., Braddick, F., Gonzalez, J. C., & Litt, J. S. (2019). Nature-based social prescribing in urban settings to improve social connectedness and mental well-being: A review. *Current Environmental Health Reports, 6*(4), 297–308. https://doi.org/10.1007/s40572-019-00251-7

Maddock, A., McCusker, P., Blair, C., & Roulston, A. (2022). The mindfulness-based social work and self-care programme: A mixed methods evaluation study. *The British Journal of Social Work, 52*(5), 2760–2777.

Natural England. (2017). *Good practice in social prescribing for mental health: The role of nature-based care.* [Report]. Natural England.

NHS England. (n.d.). *Green social prescribing.* Retrieved October 17, 2025, from https://www.england.nhs.uk/personalisedcare/social-prescribing/green-social-prescribing/

Noble, C. (2016). Green social work—The next frontier for action. *Social Alternatives, 35*(4), 14–19. Retrieved from https://www.proquest.com/scholarly-journals/green-social-work-next-frontier-action/docview/1918321298/se-2

Närhi, K. (2004). *The eco-social approach in social work and the challenges to the expertise of social work.* University of Jyväskylä.

Obeng, J. K., Kangas, K., Stamm, I., & Tolvanen, A. (2023). Promoting sustainable well-being through nature-based interventions for young people in precarious situations: Implications for social work. A systematic review.*Journal of Happiness Studies, 24*(8), 2881–2911. https://doi.org/10.1007/s10902-023-00683-x

Rapeli, M. (2018). Social capital in social work disaster preparedness plans: The case of Finland. *International Social Work, 61*(6), 1054–1066. https://doi.org/10.1177/0020872817695643

Slesinski, S. C., Matthies-Wiesler, F., Breitner, S., Gussmann, G., Schneider, A., et al. (2025). *Social inequalities in exposure to heat stress and related adaptive capacity: A systematic review.* Environmental Research Letters, 20, Article 033005. https://doi.org/10.1088/1748-9326/20/033005

United Nations Environment Programme. (2021). *Beating the heat: A sustainable cooling handbook for cities.* UNEP. https://www.unep.org/resources/report/beating-heat-sustainable-cooling-handbook-cities

Wilson, E. O. (1984). *Biophilia.* Harvard University Press.

World Health Organization. (2008). *Heat–health action plans: Guidance.* WHO Regional Office for Europe. https://www.who.int/publications/i/item/9789289071918

World Health Organization. (2024, May 28). *Climate change, heat and health.* https://www.who.int/news-room/fact-sheets/detail/climate-change-heat-and-health

World Meteorological Organization. (2025). *State of the Global Climate 2024* (WMO-No. 1368). WMO.

Chapter 5
Risk Assessment in Practice

Introduction

Social workers are used to risk, given that risk assessments are a fundamental part of our role. However, our assessments are traditionally things such as risk of physical, emotional, or sexual harm in relation to children; risk of physical injury and safety risks in relation to older adults and so on. Climate change impacts as discussed previously increases the risk in all areas of vulnerability (Thomas et al., 2019). Consequently, the thresholds currently used in risk assessment may need to be revised downward to adequately capture emerging risks. However, a coinciding increase in those needing a service, may cause an overwhelm of social service capacity—this in turn may mean a radical change in our assessment of who is the most vulnerable (Curtis et al., 2017). Those who in time of stability receive a service, may in times of crisis not be able to due to lack of workers; financial resources and social work triage of cases to only the most acute (Dominelli, 2015). So what can we do? If social work offices become inundated with those wanting help, how can we manage our risk assessments to get the right people the right help at the right time?

Firstly, we will need to identify who might already be vulnerable to different types of climate impact, such as floods; wildfires, etc., due to their location, limited adaptive capacity, and intersecting vulnerabilities. Additionally, we also need to have pre-emptive awareness of which groups are likely to move into a state of vulnerability resulting from the ways climate change will produce multi-dimensional and intersecting vulnerabilities that are not always predictable. Some individuals who are secure in current systems for example may become exposed when infrastructures and everyday services fail (electricity, transportation, and stocked supermarkets) and skills prove critical. In this way, climate impacts reveal that vulnerability is related to systems in which people live beyond what is already the case.

Paradoxically, those who are currently seen as vulnerable may demonstrate significant resilience in climate-related impacts. People with lived experience of hardship often have established informal support networks, such as neighbors who share food,

S. Rost, *Climate Resilience Guide for Social Workers*,
SpringerBriefs in Social Work, https://doi.org/10.1007/978-3-032-24604-2_5

or community ties that activate quickly in a crisis. An older person living alone might appear at risk on paper, but in practice may have extensive experience in self-sufficiency—knowing how to light fires, use alternative communication like ham radios, and maintain a stocked cupboard of tinned goods. Climate vulnerability, then, must be assessed with attention to both systemic dependencies and hidden capacities, rather than relying solely on surface traditional indicators which may become out of date.

Secondly, make yourself aware of additional resources and services that are available in your area to respond to events such as heatwaves, flooding, and other climate-related disruptions. This awareness is not only about knowing what exists, but also about understanding how accessible these services are to the people you work with. For example, during heatwaves, local authorities or NGOs may set up cooling centers, but these may be out of reach for elderly people without transport, or for families with young children who cannot easily travel. As a social worker, mapping out both the existence of resources and the barriers to accessing them can help you anticipate gaps in provision.

If you notice that there are limited or inadequate resources, you can advocate for your local office to implement new measures. These might include offering air-conditioned or cooled community spaces during heatwaves and ensuring water stations are available in central, accessible locations. In the winter, the same principle applies: community buildings or offices can serve as heated safe spaces with warm drinks, blankets, and access to emergency generators to provide electricity for those at risk during prolonged power outages.

In addition to these space-based responses, it is also important to consider direct, mobile support. While food banks and charities provide assistance, social workers are often at the frontline of urgent needs as well. Having a supply of non-perishable food items, bottled water, and hygiene products available at the office or in outreach vehicles can allow you to respond immediately to families or individuals facing sudden shortages during or after a crisis. This can be lifesaving for people who are unable to wait for formal distribution systems to activate. Food has often been at the forefront of social work support, and this has been diluted over the years in many countries—redirected to churches, foodbanks, and charities. That can still be our way of supporting the most vulnerable, given it's an immediate need.

It may also be worth thinking beyond physical resources to services such as emergency mental health support, helplines, and volunteer networks that can be mobilized during extreme events. The emotional and psychological toll of climate-related disasters is significant, and ensuring that clients know where to turn for immediate support can help reduce longer-term impacts.

While many of these initiatives may already exist in some form in your local area, the increasing likelihood and intensity of extreme weather events means they will need to be scaled up, made more robust, and integrated into everyday planning. As social workers, being proactive in identifying what is missing, advocating for additional support, and ensuring that vulnerable groups are not overlooked will be central to building resilient communities in the face of climate disruption.

So in this sense you are now having to reassess what vulnerability is, which may now vary according to which climate impact affects the population you work with. You also need to assess your provisions, your supports that you have available currently, with a view to whether they will need to be expanded or developed in the near future. How will your local social work office operate if floods not only inundate the homes of your clients, but that of you and your colleagues as well. Risk assessment therefore becomes much more holistic, encompassing workers also, and in so doing, you can assess what is realistically feasible in a crisis.

Here I give some generalized methods of risk assessing at group level and an individual level.

Social Workers as a Group, Part of a Local Office or at Municipal Level

Social workers are not islands and usually work in teams or in relation to their immediate work rules and guidelines according to their area. What this means is that they can use that shared knowledge to identify individuals and groups vulnerable to climate impacts in their communities by combining their knowledge of local contexts with structured vulnerability assessments. Team up with schools, health services, criminal justice, and addictions services to assess risk of climate impacts across services. Here's a multi-step approach you can use:

Use Existing Data and Knowledge

Review local records: Use data from municipal agencies, public health records, emergency services, or NGOs to identify individuals already flagged as vulnerable (e.g., elderly, chronically ill, low-income, undocumented migrants). Integrate this knowledge into your existing social work protection meetings.

Tap into local networks: Social workers often have informal knowledge about their community. Use this.

Apply a Climate Vulnerability Framework

Use a structured tool or framework to assess:

Exposure: Who lives in flood zones, heat islands, drought-prone areas, or food deserts?

Sensitivity: Who is more likely to be harmed? (e.g., children, people with disabilities, people with chronic illnesses, isolated elderly).

Adaptive capacity: Who has fewer resources or support systems to adapt or respond? (e.g., Single parents with no family networks, migrants without legal protection, among others).

Example Tools:

- **Climate Vulnerability and Capacity Analysis (CVCA)** (CARE, n.d.)
- **Social Vulnerability Index (SoVI)**—adapted to local contexts (Hazards and Vulnerability Research Institute, 2022)
- **Livelihood Vulnerability Index (LVI)**—simplified for local use (Hahn et al., 2009)

Each of these offers specific guides on how to carry out these assessments focusing on vulnerabilities and barriers that may exist to accessing support. This is something best not done by an individual social worker but as a group, area, with municipality involvement.

Social Workers as Individuals

What you as a social worker can do in practice is be an observer of risk. This builds on what we do already but requires another awareness of how climate effects may impact on individuals, groups, and families.

Here you find a checklist as a general guide in order to assess climate impact risks on your home visits:

1. **Heatwaves and Temperature Extremes**

Look Out for:

- No access to fans, air conditioning, or cool shelter
- People wearing heavy clothes during hot weather (due to cognitive issues or poverty)
- Signs of dehydration or heat stress (confusion, fatigue, dry mouth)
- Older adults or disabled individuals living on upper floors without lifts
- People confined to small apartments with poor ventilation
- Outdoor workers
- Children not adequately protected from heat

2. **Flooding, Storms, and Rainfall**

Look Out for:

- Homes located in low-lying areas, near rivers, or with obvious drainage problems
- People with limited mobility or no vehicle access
- Poor-quality housing: mold, leaky roofs, rising damp
- Households relying on electric-powered medical equipment or refrigerated medication (at risk in power outages)

3. **Cold Weather and Snow**

Look Out for:

- No heating or very low indoor temperatures
- People using ovens or unsafe heaters to warm homes
- People with limited warm clothing
- Difficulty affording heating bills (energy poverty)
- Isolation in rural areas or blocked-in homes
- Moisture on windows and walls
- Social isolation

4. **Food Insecurity and Supply Disruption**

Look Out for:

- People already reliant on food banks or emergency aid
- Lack of food storage (no fridge/freezer or broken appliances)
- Empty cupboards and fridges
- Individuals who live far from shops with no transport
- Signs of weight loss, malnutrition, or rationing food for children or adults
- People giving limited descriptions of what they have eaten that day
- People using food money for other bills and expenses

5. **Social Isolation**

Look Out for:

- Individuals with no one to check on them during extreme weather
- Undocumented migrants or people afraid to contact authorities
- Those potentially affected by breakdowns in community services (transport, home care, meal delivery)
- Those with language or cognitive difficulties in understanding local emergency systems

Scenario 1: Heatwave—At-Risk Family with Young Children

Names: Sandra (mother, 32), Johan (father, 34), Leo (8), and Ava (2)Setting: Home visit during early summer; heatwavealert issued for the coming week

Location: Inner-city apartment block, third floor, no air conditioning or cooling system

Observations:

- The apartment feels stiflingly hot, with poor ventilation and no visible fans or cooling options. Windows are shut, and heavy curtains trap heat.
- The parents appear tired and irritable.

- There are prior child protection concerns: previously, Sandra and Johan left 8-year-old Leo to care for toddler Ava alone while they went out. This has happened more than once.
- Although they were warned not to repeat this, Leo quietly admits that it "sometimes still happens."
- Both parents have known substance misuse issues and periods of emotional instability. They say they've "cut down drinking," but there are signs of relapse.
- The family receives minimal social support, and they are not in regular contact with extended family or neighbors.

Current Context:

- A severe heatwave is forecast for the coming week, with daytime temperatures above 32 °C and high night-time heat retention in urban apartments.
- National and local heat health advisories have been issued, especially targeting vulnerable groups, including children and those in poor housing.

Key Risks to Consider

- Young children are especially vulnerable to heat exhaustion, dehydration, and heatstroke (Centers for Disease Control & Prevention, 2022)
- Apartments may reach dangerous indoor temperatures, especially overnight (Porritt et al., 2012)
- Parents may leave children unsupervised during high-heat periods (United Nations Children's Fund, 2023)
- Neglect risk may increase if caregivers are impaired by substances or seek to leave the apartment (Ranson, 2014)
- Children may not have access to hydration, shade, or cooling (World Health Organization, 2025)
- Family may be unaware of heatwave dangers or how to protect children (United States Environmental Protection Agency, 2021)
- Urban apartments with no green space or outdoor shade increases exposure (Stone et al., 2012)

Potential Supports to Lower Risk

- Immediate safety plan: Reinforce supervision requirements and clarify that children must not be left alone-especially during high heat.
- Provide fans or cooling equipment through emergency aid or local charity partnerships.
- Check for community cooling centers or shaded public spaces and help the family make a plan to use them safely.
- Link family with intensive parenting support services, especially over the summer period
- Coordinate with addiction services for support and monitoring during the heatwave.

- Provide visual or easy-to-understand heat safety guidance, including hydration tips, signs of heat exhaustion, and how to keep rooms cooler. Leo can also be included in knowledge sharing.
- Schedule frequent check-ins or home visits during the heatwave period.
- Explore emergency respite or temporary day placements if the home environment becomes unsafe for the children, including relatives.

Scenario 2: Flooding—Elderly Person Dependent on Homecare

Name: Ingrid, 84

Setting: Call received from a concerned neighbor following flash flooding.

Location: Ground floor apartment in a low-lying residential neighborhood near a river.

Background:

- Ingrid is an elderly woman living alone with limited mobility due to arthritis and a history of strokes. She uses a walker and relies on daily homecare support for cooking, bathing, and medication management.
- She receives pre-prepared meal deliveries once a day and does not cook for herself.
- Ingrid has no immediate family nearby; her closest relatives live several hours away.
- After sudden and intense rainfall, the area has experienced flash flooding. Streets are partially submerged, and homecare workers are unable to reach her.
- A neighbor reports rising water near the building and says Ingrid has not been seen that day. The neighbor does not have a key, and Ingrid's emergency contact information appears outdated.

Key Risks to Consider

- Missed medication and inability to eat or drink safely (Ochi et al., 2014)
- Risk of floodwater entering the apartment, leading to health hazards or injury (European Climate and Health Observatory, 2025)
- Electrical outages could affect lighting, heating/cooling, and medical equipment (Tsagkaris et al., 2023)
- Social isolation may cause panic, disorientation, or confusion (Mehta et al., 2024)
- Ingrid has no immediate support network and cannot leave independently (First Option, 2025)

Who Can You Contact to Ensure Ingrid's Safety?

- Local emergency services (fire/police/rescue) for welfare check or safe evacuation.
- Adult social services or homecare coordination teams for emergency follow-up.
- Local disaster response or emergency management agencies.

- Non-profit or volunteer organizations that provide community emergency response (e.g., Red Cross, faith-based networks).
- Meals-on-wheels providers or other community food programs for alternative food delivery.
- Neighbors or building management who may be able to safely check in or offer temporary support.
- Landlord or housing provider to assess the safety of the apartment and assist with access.

Preventive Measures to Put in Place in Advance

- Work with Ingrid to create a personal emergency preparedness plan, including:
- A stockpile of shelf-stable food and bottled water
- A week's supply of essential medications in waterproof containers
- Flashlight, batteries, and emergency contact list
- Ensure backup communication options (charged mobile phone, power bank)
- Coordinate with homecare providers to identify backup staff or support options
- Create a check-in network: trusted neighbors or community volunteers who can visit or call during weather events
- Provide clear, accessible information about flood risk, evacuation plans, and safety tips tailored to older adults

References

CARE. (n.d.). *Climate Vulnerability and Capacity Analysis (CVCA).* https://careclimatechange.org/cvca/CARE Climate Change

Centers for Disease Control and Prevention. (2022). *Extreme heat and children.* U.S. Department of Health & Human Services. https://www.cdc.gov/disasters/extremeheat/children.html

Curtis, S., Fair, A., Wistow, J., et al. (2017). Impact of extreme weather events and climate change for health and social care systems. *Environmental Health, 16*(Suppl 1), 128. https://doi.org/10.1186/s12940-017-0324-3

Dominelli, L. (2015). The opportunities and challenges of social work in disaster situations. *International Journal of Social Welfare, 24*(4), 414–424. https://doi.org/10.1111/ijsw.12143

European Climate and Health Observatory. (2025, March 17). *Flooding—Health effects.* https://climate-adapt.eea.europa.eu/en/observatory/evidence/health-effects/flooding

First Option. (2025). *Personal Emergency Evacuation Plans (PEEPs).* https://data-firstoption.group/peeps/ (explains person-specific evacuation planning for people unable to self-evacuate).

Hahn, M. B., Riederer, A. M., & Foster, S. O. (2009). The Livelihood Vulnerability Index: A pragmatic approach to assessing risks from climate variability and change—A case study in Mozambique. *Global Environmental Change, 19*(1), 74–88. https://doi.org/10.1016/j.gloenvcha.2008.11.002

Hazards & Vulnerability Research Institute. (2022). *Social Vulnerability Index (SoVI®): Methodology and limitations.* https://www.vulnerabilitymap.org/Portals/0/Files/Social%20Vulnerability%20Index%20SoVI%20-%20Methodology%20and%20Limitations.pdf

Mehta, M. M., McGhee, S., Chou, J., & Jeste, D. V. (2024). Climate change and aging: Implications for psychiatric care. *Focus (American Psychiatric Association), 22*(3), 260–270. https://pmc.ncbi.nlm.nih.gov/articles/PMC11384634/

Ochi, S., Hodgson, S., Landeg, O., Mayner, L., & Murray, V. (2014). Disaster-driven evacuation and medication loss: A systematic literature review. *PLOS Currents Disasters, 6*. https://doi.org/10.1371/currents.dis.fa417630b566a0c7dfdbf945910edd96

Porritt, S. M., Cropper, P. C., Shao, L., & Goodier, C. I. (2012). Ranking of interventions to reduce dwelling overheating during heat waves. *Energy and Buildings, 55*, 16–27. https://doi.org/10.1016/j.enbuild.2012.01.043

Ranson, M. (2014). Crime, weather, and climate change. *Journal of Environmental Economics and Management, 67*(3), 274–302. https://doi.org/10.1016/j.jeem.2013.11.008

Stone, B., Vargo, J., & Habeeb, D. (2012). Managing climate change in cities: Will climate action plans work? *Landscape and Urban Planning, 107*(3), 263–271. https://doi.org/10.1016/j.landurbplan.2012.05.014

Thomas, K., Hardy, R. D., Lazrus, H., Mendez, M., Orlove, B., Rivera-Collazo, I., Roberts, J. T., Rockman, M., Warner, B. P., & Winthrop, R. (2019). Explaining differential vulnerability to climate change: A social science review. *Wires Climate Change, 10*(2), Article e565. https://doi.org/10.1002/wcc.565

Tsagkaris, C., Laubsher, L., & Matiashova, L. (2023). The impact of energy shortages on health and healthcare in Europe. *Health Science Reports, 6*(3), Article e1075. https://doi.org/10.1002/hsr2.1075

United Nations Children's Fund. (2023). *Heatwave safety tips for children.* UNICEF. https://www.unicef.org/parenting/child-care/heatwave-safety-tips-children

United States Environmental Protection Agency. (2021). *Climate change and the health of children.* EPA 430-F-16-061. https://www.epa.gov/climatechange-health/climate-change-and-health-children

World Health Organization, Regional Office for Europe. (2025, June 23). *How summer heat impacts health and how to #KeepCool.* https://www.who.int/europe/news/item/23-06-2025-how-summer-heat-impacts-health-and-how-to--keepcool

Chapter 6
Food Security and Survival Skills for Social Work in Practice

Introduction

When food systems become unstable or people struggle to access food, social services are often frontline responders. Whether it's providing financial support to buy food, working with food banks, social supermarkets or soup kitchens, social workers usually know where to direct people to get food when there is none. In response to climate change and its impacts, we need to consider the consequences on those supply chains that social workersmay rely on to help support vulnerable people, and what that means for *how* we support them. Food security is at severe risk globally, due to temperature increases which lead to harvest failures, supply disruptions, soil erosion from mass agriculture, loss of pollinators and so on. At the production level, extreme weather reduces crop yields and livestock health, while global supply shocks restrict imports. These combined pressures drive food price volatility and limit availability of key staples. Rising costs strain household budgets, forcing people to make hard choices between paying their rent, heating their homes, or feeding their families. As people reduce food quantity or quality, nutrition and health outcomes become affected negatively, particularly for children, older adults, and people with chronic conditions. This is not a new phenomenon, but the cause and likely scale of it is something we have not had to deal with previously. The results could amplify psychosocial pressures, manifesting in increased anxiety and strain for individuals and families experiencing food insecurity. These intersecting impacts push more individuals and families into contact with social services, not only for food assistance but also for broader support with housing, mental health, and child protection-linking climate-driven food insecurity directly to potentially rising service demand in social work.

This means that social workers may need to think more directly about how and in what ways they can support people in obtaining food, as it may become increasingly less available either through physical supply or heightened food prices.

S. Rost, *Climate Resilience Guide for Social Workers*,
SpringerBriefs in Social Work, https://doi.org/10.1007/978-3-032-24604-2_6

Food security can be improved at various levels: individual, community, municipal, and state through planning for future shortages. Unfortunately, these plans often don't exist, or the onus is put on the individual to have sufficient supply of food to manage through emergencies (examples—UK Government, 2024; Sweden—Myndigheten för samhällsskydd och beredskap, 2024). The emphasis on individuals to carry the burden of providing for their own food supply in emergencies, neglects the very idea of those who live day to day financially. People who depend on food banks, borrowing, welfare support, and other methods of gaining enough to eat to survive. These people are not in a position to store a surplus of food when they only have enough to eat for that day. And, with prices set to continue to increase, food will become an ever less elastic asset, one that we will need to use more of our financial resources to pay for. There are already those who cannot afford it now, so the trend moving forward is likely to be increasing numbers of people struggling to be able to access the amount and type of food that is necessary for health. The four pillars of food security are as follows:

> Physical *availability* of food: Food availability addresses the 'supply side' of food security and is determined by the level of food production, stock levels and net trade.
>
> Economic and physical *access* to food: An adequate supply of food at the national or international level does not in itself guarantee household level food security. Concerns about insufficient food access have resulted in a greater policy focus on incomes, expenditure, markets and prices in achieving food security objectives.
>
> Food *utilization*: Utilization is commonly understood as the way the body makes the most of various nutrients in the food. Sufficient energy and nutrient intake by individuals are the result of good care and feeding practices, food preparation, diversity of the diet and intra-household distribution of food. Combined with good biological utilization of food consumed, this determines the nutritional status of individuals.
>
> *Stability* of the other three dimensions over time: Even if your food intake is adequate today, you are still considered to be food insecure if you have inadequate access to food on a periodic basis, risking a deterioration of your nutritional status. Adverse weather conditions, political instability, or economic factors (unemployment, rising food prices) may have an impact on your food security status.
>
> For food security objectives to be realized, all four dimensions must be fulfilled simultaneously. (World Bank, n.d.)

In addition to this, two additional food pillars are being considered as becoming increasingly relevant, of agency and sustainability (High Level Panel of Experts on Food Security and Nutrition, 2020).

While large-scale systemic changes are crucial, social workers are often best positioned to influence the first three levels—individual, community, and municipal. At these levels, a sense of agency could be supported through a number of additions to how social workers provide support and advice around accessing food. The following is based primarily on findings from a scoping review undertaken looking for individual and household adaptations to improve food security that might be possible under climate change (Rost, Under review).

Getting Your Hands on Food

Helping clients diversify their methods of acquiring food is likely to become essential. This may include things you had not even considered before, and begrudge that at this time we even have to consider. And you're not wrong, but these are the times in which we live, and so we do as social workers always do, make the best of the situation we've got. Social workers can support others by identifying and facilitating alternative methods of obtaining food.

They can direct as they do now to social supermarkets, foodbanks, and charitable organizations that are distributing food including soup kitchens/ food hubs, and so on. This is what we already do, so we can anticipate a greater demand for these moving forward in time as the climate degenerates, which means those systems are at risk of being overwhelmed. So we need to think beyond them also—not instead of, but in addition to. For example, foraging—gathering edible plants from public or wild spaces—can supplement diets when fresh produce is expensive or unavailable. A social worker could help clients identify safe, legal foraging areas or invite local experts to teach basic wild food identification skills.

- Host a community workshop on foraging and wild plant safety.
- Create a local food directory of places people can go to obtain food, such as charities, etc.
- Work with municipalities to develop mobile food delivery programs (like Meals on Wheels for climate emergencies), including for mental health day centers, temporary housing, or remote rural areas.

Growing Food

Growing food is a method of food resilience that can be encouraged at a variety of levels, from individuals growing food in pots to community gardens set up to support a larger group. Social workers can help clients start small-scale growing projects adapted to the living environment they have, whether it's a windowsill or a garden. This may not seem to be a part of social workers' remit, but if food becomes more unavailable, the ability to support your clients to meet this basic survival need would become a priority. This is in addition to the numerous benefits that interaction with nature and green spaces have on health, both physical and mental (Boetto, 2017).

Container gardening is an option for clients in urban areas or apartments, and social workers could help source free or low-cost containers, compost, and seeds through local gardening groups or charities. They can teach clients how to grow fast-producing crops like lettuce, radishes, or herbs to supplement nutritional content, as well as calorie dense staples such as potatoes for caloric intake.

For those with outdoor space, social workers could support building raised beds or setting up permaculture gardens that include perennial plants that have fewer inputs over time. This would minimize the level of time people would need to put in to support its continuation, and give a supply of food that would grow year upon year. This would reduce additional stress growing may place on people. Social workers could also help facilitate shared garden spaces, matching clients who have land but no time with others who are able-bodied and looking to garden.

Supporting learning in how to grow more climate-resilient crops such as kale, potatoes, beans, and squash—especially those with long shelf lives—could help families plan for food insecurity. On a more low-key intervention, they could promote microgreen growing indoors during colder months, which can be a useful source of nutrients.

- You could help set up a local seed library or food exchange group.
- Set up a community garden where food can be shared as well as the workload.
- Support people to obtain growing containers, growing materials such as compost, dirt, etc.
- Use growing areas around the social work building, pots that are currently filled with flowers to fill with perennial vegetables and/or fruit trees that people can use for free.

Food Storage and Preservation

Food storage and preservation have long been vital survival skills across cultures, used to carry households through seasons of scarcity. In the era of climate disruption, where food systems and supply chains are increasingly unstable these methods can play an important role in strengthening local and household food security. Social workers, particularly those working with vulnerable communities, can integrate food preservation into empowerment and resilience programs. Teaching these skills not only extends the availability of food and reduces waste, but also builds self-reliance and mutual support in the face of environmental uncertainty. Below are a number of examples of food preservation that could be utilized:

Canning

What It Is:

Canning preserves food by sealing it in airtight jars and heating it to eliminate bacteria. Two common types are water-bath canning, used for high-acid foods such as fruit, jams, and tomatoes (with added lemon juice), and pressure canning, which safely preserves low-acid foods like vegetables, beans, soups, and meats.

How social workers can help: Social workers can partner with community groups to host or coordinate introductory canning workshops, emphasizing food safety and affordability. They can distribute simple checklists or safety guides from trusted sources such as agricultural extensions or national food preservation agencies.

A source of information on this topic can be found at National Center for Home Food Preservation—Canning Guides (National Center for Home Food Preservation, n.d.).

Freezing

What It Is:

Freezing slows the growth of bacteria and preserves food quality. Vegetables are often blanched (briefly boiled) before freezing to retain texture and nutrients.

How social workers can help:

Practitioners can teach households how to label and date frozen foods for easier rotation, and share advice on using energy-efficient freezers. Encouraging freezing of surplus produce during high-yield seasons or grocery discounts helps families save money and reduce food waste.

A source of information on this topic can be found at NCHFP Freezing Guide (National Center for Home Food Preservation, n.d.).

Fermenting and Pickling

What It Is:

Fermentation uses natural bacteria to preserve foods while enhancing nutritional content. Pickling, in contrast, relies on vinegar or brine to extend shelf life and create distinct flavors.

How social workers can help:

Social workers can suggest simple, low-cost fermentation and pickling methods that require minimal equipment and can preserve vegetables long after their usual shelf life. These practices can also reconnect individuals with cultural traditions that already use such techniques.

A source of information on this topic can be found at Beginner's Guide to Fermentation—Cultures for Health; Pickling Basics—Ball Canning (Ball Mason Jars, n.d.).

Root Cellaring

What It Is:

Root cellaring makes use of naturally cool, dark, and humid environments-such as basements or underground spaces-to store vegetables like potatoes, carrots, onions, and apples without the need for electricity. This is beneficial in case of power outages, or storing food that has been grown.

How social workers can help:

Social workers can help households identify potential spaces suitable for simple root-cellar storage, or work with local organizations to establish shared cellaring facilities. They can also teach which foods keep best and how to rotate stock for long-term freshness.

A source of information on this topic can be found at Root Cellaring 101—Mother Earth News (Mother Earth News, n.d.).

Rotating Stock (First in, First Out)

What It Is:

The "First In, First Out" (FIFO) system ensures that older food is used before newer items, reducing spoilage and waste.

How social workers can help:

Social workers can discuss and encourage FIFO principles into household budgeting and emergency preparedness plans. Demonstrating simple labeling systems or storage trackers can make this routine easy to maintain, especially for larger families or shared housing.

A source of information on this topic can be found at Ready.gov—Food Storage and Rotation (Ready.Gov, N.D.).

Storage in Mylar Bags with Oxygen Absorbers

What It Is:

This long-term storage method keeps dry goods like rice, beans, flour, or oats fresh by sealing them in Mylar bags with oxygen absorbers, which slow down oxidation and pest damage.

How social workers can help:

For clients interested in preparedness or bulk purchasing, social workers can recommend safe Mylar storage practices or connect them to local preparedness groups that pool resources to reduce costs.

A source of information on this topic can be found at Mylar Bag Storage Basics—The Provident Prepper (The Provident Prepper, n.d.).

Food Sharing and Knowledge Sharing

Food sharing is something used in communities to support vulnerable groups but also to build networks and bonds across those who are isolated. While it might seem counterintuitive in times of scarcity, sharing food ensures that food is distributed more evenly and that no one is left without support. Social workers can nurture this sense of reciprocity and solidarity by organizing communal meals, food swaps, or community kitchens. Such initiatives strengthen social networks, reduce isolation, and ensure that surplus food is redirected to those who need it most.

Having knowledge of what to do in a crisis is perhaps, unsurprisingly, a useful resilience factor and this can also apply in improving food security. Knowledge of the local area, what plants grow and where combined with more pragmatic knowledge of where the social supermarket is, which charities are distributing food and who might be able to help when times are hard. All of this knowledge is something that social workers can learn and pass on to others, so that those we help can become more resilient independently of us. People typically fall back on what they know during a crisis or disaster, so social workers should no longer be holders of knowledge but distributors of it so that people can help themselves when they need it. How social workers can help:

- Establish regular food-swap events or shared food shelves.
- Coordinate meal-sharing projects in partnership with local organizations or faith groups.
- Collaborate with retailers, farms, or restaurants willing to donate surplus food for redistribution.

In case work, include clients and their children in discussion at an age-appropriate level of food security methods.

Educate yourself about indigenous or local food practices that are already in existence and ways to distribute the knowledge, through pamphlets or in person.

Guide: Arun & Chichester Food Partnership- A guide to setting up community food initiatives: Local food action toolkit (Arun & Chichester Food Partnership, 2023*)*.

Assessing for Food Insecurity

Social workers may wish to use a Rapid Food Vulnerability Screen, which is a short, five-minute tool for frontline practitioners to quickly identify households at risk of food insecurity. It asks about recent concerns over food running out, inability to afford more when supplies were exhausted, special dietary or sensory needs, the household's current ability to cook and store food safely, and barriers to accessing food such as finances, transport, mobility, language, or documentation. Together, these questions provide a snapshot of immediate food vulnerability and help guide referrals and tailored support.

A. **Rapid Food Vulnerability Screen (Client-Facing, 5 Minutes)**

1. In the past 30 days, have you been worried food would run out? (Often/Sometimes/Never)

2. In the past 30 days, has the food you bought not lasted, and you had no money to get more?
3. Have you any special dietary/texture/sensory needs?
4. Have you got the ability to cook/store food safely this week (power, water, appliances)?
5. Are there any access barriers this week (money, transport, mobility, language, documents)?

Case Study 1: Power Down, Fridge Empty

An elderly woman living alone in a rural area experiences a multi-day power outage after a summer storm. Her fridge food spoils, and she doesn't drive. She has limited savings and no emergency supplies. The nearest shop is 20 km away, and her daughter lives in another town.

Questions for Reflection:

- What immediate steps could you take to support her access to food?
- How could you help her prepare for future disruptions?
- What community partnerships could be mobilized here (e.g., mobile meals, mutual aid)?

Case Study 2: The Single Mum in the High-Rise

A single mother with two young children lives on the 5th floor of a city apartment block. She receives benefits, but rising food prices mean she often skips meals. She has no access to a garden and feels overwhelmed by trying to feed her children on a limited budget.

Questions for Reflection:

- What practical growing or preserving techniques could be adapted for her situation?
- How might you help her access affordable, healthy food?
- Could this be an opportunity to introduce her to food co-ops/social supermarkets, window gardening, or communal cooking?

Case Study 3: Community Center on a Budget

A social worker runs a drop-in space in a low-income neighborhood. Many clients experience food insecurity. The center has no kitchen but does have a yard, a few volunteers, and a small budget.

Questions for Reflection:

- How could you initiate a small food-growing project?
- What partnerships (e.g., with local farms or garden centers) might help?
- Could you involve clients in setting up food swaps or communal growing efforts?

References

Arun & Chichester Food Partnership. (2023, November). *A guide to setting up community food initiatives: Local food action toolkit*. Arun & Chichester Food Partnership. https://secure.toolkitfiles.co.uk/clients/41305/sitedata/Access_To_Food/ACFP-Food-Action-Toolkit-1-compressed.pdf

Ball® Mason Jars. (n.d.). *Canning & preserving guides*. https://www.ballmasonjars.com/canning-preserving-guides.html

Boetto, H. (2017). *An ecologically centred approach in social work: Towards transformative change (Doctoral thesis*. Charles Sturt University.

High Level Panel of Experts on Food Security and Nutrition. (2020). *Food security and nutrition: Building a global narrative towards 2030* (HLPE Report No. 15). Committee on World Food Security.

Mother Earth News. (n.d.). *Root cellaring 101: Simple methods for storing vegetables, fruits & nuts*. https://www.motherearthnews.com/real-food/root-cellaring-101-zmaz77sozgoe/

Myndigheten för samhällsskydd och beredskap. (2024). *Grunder i hemberedskap* [Basics of household preparedness]. Retrieved [today's date], from https://www.msb.se/sv/rad-till-privatpersoner/hemberedskap---preppa-for-en-vecka/grunder-i-hemberedskap/

National Center for Home Food Preservation. (n.d.). *Canning guides*. University of Georgia, College of Family and Consumer Sciences. https://nchfp.uga.edu/how/can_home.html

National Center for Home Food Preservation. (n.d.). *Freezing guide*. University of Georgia, College of Family and Consumer Sciences. https://nchfp.uga.edu/how/freeze.html

Ready.gov. (n.d.). *Food storage and rotation*. U.S. Department of Homeland Security. https://www.ready.gov/food

Rost, S. (under review). Individual and household adaptations for food security under climate change: A PRISMA-ScR global scoping review (2014–2024). Manuscript submitted for publication.

The Provident Prepper. (n.d.). *Mylar bag storage basics*. https://theprovidentprepper.org/mylar-bag-storage-basics/

UK Government. (2024). *Get prepared for emergencies*. Retrieved [Oct 30,2025], from https://prepare.campaign.gov.uk/get-prepared-for-emergencies/

World Bank. (n.d.). *What is food security? There are four dimensions*. https://www.worldbank.org/en/topic/agriculture/brief/food-security-update/what-is-food-security

Chapter 7
Practicing Social Work Amid Societal Breakdown

Introduction: Beyond the Limits of Stability

As the climate continues to degenerate, the very real possibility of societal collapse rears its ugly head to scream loudly in our ears—"and here's another thing to think about!" Just what we needed. The higher the temperature climbs the greater the risk that social systems will begin to break down, and beyond three degrees and above, industrialized societies may be on the brink (Anderson, 2011; Lynas, 2020). In earlier work (Rost, 2025), I introduced the concept of *collapsed social work*—a practice framework designed for contexts in which state infrastructure and institutional supports have disintegrated. Rather than assuming the possibility of recovery or a return to pre-collapse conditions, this model begins with the premise that collapse may be ongoing or permanent. It calls for a transfiguration of care, support, and resilience that is grounded in adaptability, mutual aid, and horizontal, community-based structures. This chapter builds on that premise, but also considers how societal collapse may affect in very practical ways the ways social workers support those in need, potentially stalling it, potentially reimagining what it might look like.

Collapse Is not New

Societal collapse does not have to mean a sudden apocalypse where everyone perishes, it is possible for societies to collapse, but the people to live on—whether through migration, reforming to smaller groups or restructuring into something new. But when we think of societal collapse brought about by climate changes, it is safe to say that the risks could be global due to the interconnected nature of our food systems, political systems and communication systems (Kemp, 2025). Collapse appears in our minds often through the narratives and stories we have heard and seen in movies and books such as Children of Men, The Road and Mad Max. These grim perspectives

S. Rost, *Climate Resilience Guide for Social Workers*,
SpringerBriefs in Social Work, https://doi.org/10.1007/978-3-032-24604-2_7

paint a picture of high levels of conflict and breakdown of food supplies, as well as any sense of social order and values. Survival of the fittest becomes the dominant premise and impulse. However, what we need to understand about collapse is that it is something that has happened many times before in human history, and will undoubtedly happen again, whether exacerbated by climate change or not. It is written as the inevitable outcome of many societies as told in the history of civilization. Luke Kemp argues (2025) that while the picture we have is of chaos and scraping for survival, collapse can in fact sometimes be beneficial for those lower in the societal hierarchies. This is because it is usually vast inequalities and unfair distribution systems that have contributed to its collapse, so those who were being squeezed, now are able to have more resources and have been seen to become at times physically healthier.

How collapse could or will occur alongside climate change and temperature increases may not have this potential positive aspect due to the negative effects on food and water supply, that may mean communities may not be able to "pull together" to improve their situation quite in the same way. Beyond a certain temperature plants will not be able to grow and disruptions to the water cycle may affect water availability (IPCC, 2022). It may also occur unevenly and intermittently, manifesting in different forms across regions and periods. Another important point to remember is that collapse in the way I describe it here is already occurring today, and has been in many parts of the world without the title of societal collapse. Many countries don't have luxuries of running water, electricity grids that are reliable or health care systems that improve health outcomes, in the same way that other countries take for granted. So the very idea of societal collapse as being a breakdown of these is through the lens of a high-income country perspective. That should show us that even in societal breakdown, social workers still would have a role to play in supporting people, but it may not be in a paid role, but one of humanitarian support to our fellow man. We can anticipate this because we can see evidence of communities helping and supporting each other even in places where the basic amenities people take for granted aren't available. The skills social workers typically have—identifying resources, networking people and supports, community building are all those we can see might be extremely valuable when stability wavers or collapses. So how, I hear you ask, can we possibly think about practicing social work in a situation where the profession may no longer exist in the form it currently does, embedded in institutions and charities with clear routes of funding and resources? Well, in response to this I would argue that we can easily imagine a doctor still continuing to use their skills to heal, even if they weren't going to be paid for it in a crisis, and so social workers can and should view their own skills as similarly unique and valuable. Understanding collapse as a *context* rather than a *catastrophe* allows for a reframing of social work's role not as a stabilizing force attempting to return society to its previous state, but as an adaptive, grounded practice that responds to emergent needs, decentralized realities, amid a long emergency of decline.

What Might the Effects Be and How Can We Help?

In the description of what collapse might look like as outlined above, we can posit some speculations on what might be affected within social systems. In this way, we can try to anticipate what social workers might be able to do in collapse conditions.

Economic Collapse

One of the primary reasons given for societal collapse is the economic collapse of a society. The financial system crashes, money falls in value, inflation skyrockets and people cannot pay their mortgages and rent. When financial institutions fail, credit markets freeze and people's jobs and stability become affected. Inflation escalation can make essential goods such as food, fuel, and medicine unaffordable. Savings evaporate, pensions become worthless, and people are unable to pay their mortgages, rent, or debts, leading to mass defaults and homelessness. As formal employment collapses, people turn to informal or barter-based economies, local exchange systems, or survival activities such as subsistence farming and scavenging. For many this would not only be a material shock but a psychological one.

Case Example:

In 2019, as Venezuela's economic and social systems collapsed, shortages of food, medicine, and essential services deepened, leaving the most vulnerable people exposed to severe hardship. In response, the International Committee of the Red Cross (ICRC, 2020) expanded its presence to provide humanitarian assistance to those affected. The ICRC aimed to maintain access to care and protection. Its support also extended to nearby islands such as Aruba, Bonaire, Curaçao, and Trinidad and Tobago, where it focused on helping detained or separated migrants and other vulnerable groups.

Electricity Failure

The electricity grid that we take for granted relies on the energy provision to generate it, maintenance of services and a large number of staff and their salaries to keep it going. Under societal collapse conditions, economic breakdown may have a cascading effect on other services. The financial resources to keep the systems going could fail, coupled with climatic impacts on electrical grids could lead to intermittent provision or blackouts. Lack of stable government oversight to ensure its continuation could potentially mean small groups and communities may need to have backup power supplies, batteries, generators or solar cells not tied to the grid. Social workers in this situation need to consider the significant effects of loss of electricity, including loss of heating, air conditioning, refrigeration of food, baby milk and medicine. They may need to work together to develop practical solutions amidst uncertainty and chaos. We can use examples from crises that have occurred to plan for what could be done.

Case Example:

Puerto Rico, 2017–2018 (months-long blackouts after Hurricane Maria): With grid power down for weeks to months, hospital and community social workers pivoted to low-tech alternatives. Many partnered with *Centros de Apoyo Mutuo* (mutual-aid centers) that emerged in schools and churches to provide meals, charging stations, information, and psychosocial support. There was sustained service disruption, but also adaptation by health/social-service personnel and mutual-aid organizers to maintain basic protection and practical support until grid power returned (Noboa-Ramos, 2023).

Internet Breakdown

We are now tied to our need for internet access for the vast majority of our information on what is going on in the world, for communication and even for accessing our finances. Many countries have moved toward cashless societies meaning our ability to access our various accounts requires the internet. Communicating via telephone is heavily internet dependent. With risks currently coming from cyber attacks, and the dependence of services being that functional societies will have financial provision to continue them, societal collapse may also impact this. Social workers would need to think about how best to communicate with each other, and with those they want to support should there be a communication blackout. It may mean you can't simply search for answers and solutions from an internet browser, and the dusty books on your shelves may be all you have. So making sure you still have books, and ones that focus on aspects of self-sufficiency and survival could be a useful preparatory measure, not just for the idea of societal collapse, but for climate impacts in general. Have a backup plan of how you might keep communication lines open. Human mesh networks (Rost, In Press) is another useful tool where networks of individuals maintain communication with each other as part of larger networks, so that information can be shared and utilized on a wider basis.

Case Example:

During the 2011 Tōhoku earthquake and tsunami in Japan, widespread power outages caused interruptions of internet and phone networks, forcing social workers to adapt quickly to offline methods of support. Despite the technological breakdown, social workers maintained essential communication, organized food and shelter support, showing that effective practice can continue through direct human connection even without digital infrastructure (Yamamura et al., 2014).

Food and Water Shortages

Alongside the climate impacts, societal breakdown would likely lead to breakdowns in the provision of food supply and if occurring on a global scale would curtail imports. Now all you have is what you can produce or have stored. The choice of food would likely therefore be significantly reduced, and we would need to supplement our diets heavily with what we as individuals, groups, communities could grow or create. I use the word create due to the limits at which growing will be realistic, but with the view that alternatives to agriculture may be possible (Denkenberger et al., 2021). Social workers may be able to utilize their skills in promoting learning in community groups about methods of creating or obtaining food. They may be able to encourage food sharing, communal kitchens and community food stores. In this way, they can use their skills to bring people together, to share our skills and knowledge as well as our resources and in this way reduce the vulnerability of all. Individuals facing food and water shortages will suffer more greatly than a community working together for the benefit of all, and it is in building those relationships and skills that social workers can make a difference for the better. More concrete examples of how social workers can improve food security are given in Chap. 6.

Case Example:

After Typhoon Haiyan (Yolanda) struck the Philippines in 2013, social workers played a key role in ensuring access to food and water for affected communities and supporting those who had experienced trauma, alongside often sharing that trauma as victims themselves (International Federation of Social Workers, 2014). The Department of Social Welfare and Development (DSWD) coordinated large-scale distribution of relief goods, while local social workers worked alongside volunteers, humanitarian organizations, and local leaders to identify vulnerable households and maintain social support networks. Their actions demonstrated how social work practice can adapt to crisis conditions-combining emergency logistics, advocacy, and care to meet essential survival needs when formal systems are overwhelmed (International Federation of Social Workers, 2014).

Health Care Breakdown

Our health systems depend on a number of key resources—hospitals/health centers; doctors, nurses, and other medical staff; medicines ranging from painkillers to chemotherapy and anesthesia. Health care breakdown amidst societal breakdown would likely compromise all of these areas. Hospital facilities are dependent on electricity, heating, and the ability to sterilize instruments to prevent infection. These use both electricity and certain chemicals and so without these, infection risk goes up and alternatives would need to be used. If anesthesia was limited or not available, many surgeries would become impossible without killing the patient, meaning more people would suffer and die from conditions that are not treatable under those conditions. Medicines have been the key to increasing life expectancy in the twentieth and twenty-first century, and if these became limited or not available, we would likely use herbs and plant-based treatments that are only suitable for mild illness. So what can we do as social workers? We can support our medical colleagues, who will continue to try to help those who are injured or in pain. We can learn basic skills in first aid so that they can concentrate on emergency interventions. We can provide emotional support to those that are suffering and their families, just as we do now in hospital social work.

Case Example:

During the flooding caused by Hurricane Harvey in 2017, social workers worked alongside medical staff in maintaining the functioning of overwhelmed health services. Social workers supported those facing emotional trauma of the disaster, helping those displaced to access shelters and supplies (Crawford, 2021).

Domestic Violence

As climate change warms the planet, there is an association with increases in incidents of domestic violence (Choi et al., 2024). Crisis and conflict appear to increase the risk to women of violence and assault due in part to lack of law enforcement available (Thurston et al., 2024), and we may predict that in times of societal breakdown, women may be more vulnerable in this way. Social workers would need to adopt more adaptive, community-based approaches to safety. Refuges would perhaps need to become local and discreet, taking the form of trusted homes, women's collectives, or faith-based spaces embedded within communities rather than formal institutions. Social workers can help create and map these safe networks, establish quiet communication systems, and prioritize immediate survival needs. Women would need to support each other through mutual aid and social bonds. This reimagined practice of social work focused on solidarity and local resilience, means that we would still have the ability to protect those who are vulnerable.

Case Example:

Following Hurricane Maria in 2017, rates of domestic violence in Puerto Rico increased sharply, with reported homicides linked to intimate partner violence nearly doubling and requests for survivor services rising by over 60% (Cheng, 2024). Social-work-led organizations such as ESCAPE played a central role in responding to this crisis, actively shielding women from harm as well as having to manage such a significant increase in cases and in the severity of these (González-Ramírez, 2018).

Education

When social systems such as governance, food, health, and welfare begin to fragment, education often remains one of the last community functions capable of transmitting stability. Educational spaces also provide opportunities for mutual support and psychosocial recovery, particularly for children and young people affected by instability and loss. Beyond academic learning, education helps communities reinterpret crises, transforming fear and uncertainty into shared understanding and adaptive capacity, and thereby preventing hopelessness and apathy. Within a framework of social work resilience, this highlights the importance of recognizing schools, community learning circles, and peer-education networks as essential forms of social protection in their own right.

Case Example:

After the tsunami in Indonesia, charities focused on supporting children to receive some sort of learning opportunities within 30 days. It was thought that given the trauma they had experienced, returning to school provided a sense of stability and normality to support them emotionally as well as allowing their parents space to regroup and rebuild their lives (Briggs, 2018).

Mental Health

I talk about mental health in more detail in the next chapter, as climate change and the fears surrounding it, have been seen to increase mental health issues such as eco-anxiety, eco-depression and eco-anger (Stanley et al., 2021). To hypothesize how this may play out in a situation of a societal collapse we can use some speculation alongside some real-life examples of how people cope in high stress survival events, wars, and disasters. People in the immediate aftermath of an environmental disaster can experience acute stress and anxiety, but can benefit from resilience strategies and support to communities (Makwana, 2019). Mental health disorders such as depression, post-traumatic stress disorder, and stress related distress have been shown to be evident in these situations, perhaps unsurprisingly (North & Pfefferbaum, 2013). According to the World Health Organization, 22% who have experienced war in the last ten years end up experiencing mental health conditions including depression, anxiety, PTSD, bipolar or schizophrenia (World Health Organization, 2024). The WHO also advises that people with preexisting conditions may be especially vulnerable in these situations. So we can surmise that with breakdowns in social systems and elements of disaster, that social workers would need to be able to quickly assess and put in place psychological intervention. This can include psychological first aid, with key suggestions of: (1) Look; (2) Listen; (3) Link—linking to appropriate services (WHO, 2024). If these services are fragmented or not available, you may need to be that service, or create a network of those trained in counseling or psychology who may offer support to those in crisis. You must remember that you too are not existing in a vacuum but also a human being who may be experiencing the mental health impacts of your own experiences. Experiences and burdens must be shared, no one can be an island taking on the weight of a collapsing world alone. Friends, family, colleagues, community will all fare better if we are acting together and in support of one another.

All of the aforementioned case examples include support for the mental health of those they help.

Conclusion

Collapse, in this view, is not an endpoint, but a condition under which new forms of ethical, relational, and localized social work may emerge. In conditions of societal breakdown, social workers may need to redefine their practice entirely. In urban areas affected by economic collapse and infrastructure failure, community survival could depend on converting abandoned public spaces into shared hubs for food, childcare, and shelter. Professional hierarchies would give way to networks of trust, with former social workers acting as facilitators, mediators, and coordinators of collective decision-making. Mental health care would shift toward mutual support and social bonding.

In rural or isolated settings, the emphasis would move to sustaining life through cooperation and ecological knowledge. Practitioners embedded in these communities might focus on food sovereignty, low-tech healthcare, and shared caregiving, drawing on local resources and social cohesion rather than formal institutions.

Where fragments of the state still operate, social workers could become vital intermediaries-translating between official systems and community realities. They might help people access scarce aid, repurpose public spaces, and organize local responses when government services falter.

Across all these contexts, the essence of social work—protection, coordination, equity, and collective resilience—remains, but the work itself becomes grounded in adaptability and sustaining some level of care when formal systems no longer function.

Reflection and Discussion Questions

1. How does understanding collapse as an ongoing context, rather than a one-time catastrophe, change the way we think about the role of social work?
2. What ethical responsibilities do social workers have when the systems they rely on begin to fail?
3. In what ways can mutual aid and community-based support serve as alternatives to formal services during systemic breakdowns?
4. What skills or mindsets might social workers need to develop to remain effective in a collapsing or post-collapse society?
5. How does planning for collapse challenge or complement the traditional goals of social work?

References

Anderson, K. (2011, November 29). *Going beyond dangerous climate change: Exploring the void between rhetoric and reality in reducing carbon emissions*. Lecture presented at the Royal Society, London, UK. https://kevinanderson.info

Briggs, B. (2018, December 13). *Safe schools: Getting children back into education after natural disaster*. Theirworld. https://theirworld.org/news/safe-schools-getting-children-back-into-education-after-natural-disaster/

Cheng, M. (2024, December 12). *Gender-based violence: The unseen toll of hurricanes*. Think Global Health. https://www.thinkglobalhealth.org/article/gender-based-violence-unseen-toll-hurricanes

Choi, H. M., Heo, S., Foo, D., Song, Y., Stewart, R., Son, J., & Bell, M. L. (2024). Temperature, crime, and violence: A systematic review and meta-analysis. *Environmental Health Perspectives, 132*(10), Article 106001. https://doi.org/10.1289/EHP14300

Crawford, C. A. (2021). Studying social workers' roles in natural disasters during a global pandemic: What can we learn? *Qualitative Social Work : QSW : Research and Practice, 20*(1–2), 456–462. https://doi.org/10.1177/1473325020973449

Denkenberger, D., Cole, D., & Griswold, M., et al. (2021). Resilient foods for global catastrophic risks. *Frontiers in Sustainable Food Systems, 5*, Article 673081. https://doi.org/10.3389/fsufs.2021.673081

González-Ramírez, A. (2018, September 19). *After Hurricane Maria, a hidden crisis of violence against women in Puerto Rico*. Refinery29. https://www.refinery29.com/en-us/2018/09/210051/domestic-violence-puerto-rico-hurricane-maria-effects-anniversary

International Committee of the Red Cross. (2020, May 14). *Venezuela: The ICRC is stepping up its humanitarian response to the needs of the most vulnerable*. International Committee of the Red Cross. https://www.icrc.org/en/document/venezuela-icrc-stepping-its-humanitarian-response-needs-most-vulnerable

International Federation of Social Workers. (2014, January 26). *Social work in the typhoon affected areas of the Philippines: A call to the international social work community.* https://www.ifsw.org/social-work-in-the-typhoon-affected-areas-of-the-philippines-a-call-to-the-international-social-work-community/

IPCC. (2022). *Climate change 2022: Impacts, adaptation and vulnerability. contribution of working group II to the sixth assessment report of the intergovernmental panel on climate change* (H.-O. Pörtner et al., Eds.). Cambridge University Press. https://www.ipcc.ch/report/ar6/wg2/

Kemp, L. (2025). *Goliath's curse: The history and future of societal collapse*. Penguin Books Ltd.

Lynas, M. (2020). *Our final warning: Six degrees of climate emergency*. HarperCollins.

Makwana, N. (2019). Disaster and its impact on mental health: A narrative review. *Journal of Family Medicine and Primary Care, 8*(10), 3090–3095. https://doi.org/10.4103/jfmpc.jfmpc_893_19

Noboa-Ramos, C., Almodóvar-Díaz, Y., Fernández-Repollet, E., & Joshipura, K. (2023). Healthcare and social organizations' disaster preparedness, response, and recovery experience: Lessons learned from hurricanes Irma and Maria. *Disaster Medicine and Public Health Preparedness, 17*, Article e306. https://doi.org/10.1017/dmp.2022.272PMC+2PubMed+2

North, C. S., & Pfefferbaum, B. (2013). Mental health response to community disasters: A systematic review. *JAMA, 310*(5), 507–518. https://doi.org/10.1001/jama.2013.107799

Rodríguez-Díaz, C. E., Garriga-López, A. M., Malavé-Rivera, S. M., & Vargas-Molina, R. L. (2018). Challenges and responses to public health emergencies on the island of Puerto Rico after Hurricane Maria: Implications for social work practice. *Journal of Social Work Values and Ethics, 15*(2), 37–46. https://jswve.org/download/15-2/articles15-2/JSWVE-15-2-Challenges-and-responses-to-public-health-emergencies-on-the-island-of-Puerto-Rico-after-Hurricane-Maria.pdf

Rost, S. (2025). *Collapsed social work: A theoretical framework for adapting to societal collapse*. Advance online publication. https://doi.org/10.1332/20498608Y2025D000000086

Rost, S. (in press). *Human mesh networks: Communicating social work research during collapse conditions*. Social Work.

Stanley, S. K., Hogg, T. L., Leviston, Z., & Walker, I. (2021). From anger to action: Differential impacts of eco-anxiety, eco-depression, and eco-anger on climate action and wellbeing. *Journal of Climate Change and Health, 1*, Article 100003. https://doi.org/10.1016/j.joclim.2021.100003

Thurston, A. M., Stöckl, H., & Ranganathan, M. (2021). Natural hazards, disasters and violence against women and girls: A global mixed-methods systematic review. *BMJ Global Health, 6*(4), Article e004377. https://doi.org/10.1136/bmjgh-2020-004377

United Nations Population Fund. (2025, February 26). *Mental health support teams in Ukraine offer a lifeline in a warzone*. UNFPA. https://www.unfpa.org/news/mental-health-support-teams-ukraine-offer-lifeline-warzone

World Health Organization. (2024, May 15). *Mental health in emergencies: Fact sheet*. https://www.who.int/news-room/fact-sheets/detail/mental-health-in-emergencies

Yamamura, H., Kaneda, K., & Mizobata, Y. (2014). Communication problems after the Great East Japan Earthquake of 2011. *Disaster Medicine and Public Health Preparedness, 8*(4), 293–296. https://doi.org/10.1017/dmp.2014.49

Chapter 8
Maintaining Mental Health Amid Ongoing Eco-Anxiety and Climate Grief: Being the Rock in the Stream

Introduction

Social workers are not immune to the emotional toll of climate collapse. This chapter offers tools and frameworks to manage burnout, eco-anxiety, and climate grief—both personally and in those we support. It draws on trauma-informed practices and collective care approaches, encouraging professionals to become resilient presences: "rocks in the stream" amid rising distress. Case studies help translate these ideas into practice.

In the countless inspirational quotes of endurance, one poignant example says, "In the confrontation between the stream and the rock, the stream always wins not through strength but through perseverance" (Brown, n.d.). In response to climate change I think the reverse might be true. Social workers have the power to be the rock in the stream, with climate changes flowing past, staying strong so that others may find footing. Rocks do not stand alone in the stream—they're supported by other rocks around them, fortified by pressure and proximity. And while eventually the stream may be beyond their power or control, they can remain united and stronger together.

Mental health in an age of climate chaos and disaster takes on a whole new meaning, depth, and scope. With the world lurching from one crisis to another—pandemics, wildfires, flooding, heatwaves, war, economic instability, how does this sustained state of instability affect our mental health? Are we all quietly unraveling, trying to cope with a sense of ambient doom while juggling the responsibilities of daily life and work? I know I am. In moments of clarity or breakdown, it often feels like what we need is a species-wide group hug—an acknowledgment that *this is not normal*, that *we* are not maladapted for struggling in a collapsing system; the system itself is maladapted to the reality we now inhabit.

The psychological impacts of climate change are increasingly being recognized not only as important but as urgent. Doherty and Clayton (2011) outline these impacts in three interlinked categories: direct, indirect, and psychosocial.

S. Rost, *Climate Resilience Guide for Social Workers*,
SpringerBriefs in Social Work, https://doi.org/10.1007/978-3-032-24604-2_8

- Direct impacts stem from experiencing climate-related disasters such as fires, floods, droughts, hurricanes, or extreme heat, which can result in trauma and loss of family members or belongings and homes. These traumatic experiences can lead to conditions such as post-traumatic stress disorder (PTSD), anxiety disorders, and depression.
- Indirect impacts emerge from the disruptions that follow: economic instability, food and water insecurity, a need to migrate, and social unrest. These can degrade mental wellbeing gradually, particularly in communities already experiencing intersectional vulnerability.
- Psychosocial impacts are perhaps the most widespread yet least visible. They include chronic worry, existential dread, grief, and a feeling of helplessness that accumulates over time through exposure to the worsening state of the planet. This can come through pervasive media coverage of negative events as well as personal experience of ongoing environmental changes. These impacts are especially pronounced among those who have an understanding of the science and what the possible outcomes are, but feel unable to meaningfully alter its trajectory. This can lead to a sense of helplessness or despair (Doherty & Clayton, 2011).

Importantly, climate-related mental health stressors do not exist in a vacuum. They interact with and exacerbate pre-existing social inequalities. Those already experiencing health disparities, poverty, racism, gender inequalities, or mental illness are more vulnerable to the psychological consequences of climate disruption (Berry et al., 2010; White et al., 2023; World Health Organization, 2023). In contrast, those who benefit from relative geopolitical stability, wealth, and access to mental health support, especially young, wealthy, white men in the Global North, face fewer mental and physical risks from climate change in the short term (Chance et al., 2023; Oxfam International, 2023).

The concept of eco-anxiety has gained attention in recent years, particularly as affecting young people (Stanley et al., 2021). It describes persistent worry and feelings of doom about climate change and its potential impacts. This is related to its brothers—eco-depression and eco-anger—which are also emerging as recognized responses to climate changes. The term solastalgia, coined by philosopher Glenn Albrecht, captures a grief and longing experienced when familiar places are altered or lost due to environmental change, akin to feeling homesick when you're already home (Albrecht, 2005).

The emotional toll is not simply our thoughts and feelings, it also affects behavior and social relationships and consequently affects health systems that treat people experiencing these. These feelings can manifest in somatic disorders, increased substance use, and even suicidality (Burke et al., 2018). For example, studies link extreme heat events to physical illness but also to rising rates of domestic violence (Mulambo et al., 2025), self-harm (Berry et al., 2018) and even insomnia (Obradovich et al., 2024). Chronic exposure to climate-related stressors could then erode people's ability to cope and a sense of hope for the future-all key components of psychological resilience.

In this context, it is vital to note that emotional responses influence what actions we then take. While grief and anxiety are valid responses, they can sometimes become paralyzing, leading to avoidance or chronic despair (Hickman et al., 2021). This emotional overwhelm, particularly when unsupported by friends, family, and a sense of agency to make a difference, can diminish both how you feel and whether you take any positive action.

In contrast, eco-anger, where you focus your emotional response against a perceived injustice, has been found to correlate positively with climate activism and greater psychological resilience (Hickman et al., 2021). Anger can clarify boundaries, give you focus on something larger than yourself and also a target against which to react. So in that sense, not all distress is disempowering. Climate anxiety can lead to a continuum of coping responses—while some people may feel overwhelmed or stuck, others may channel their anger into meaningful action, and such engagement can actually support more adaptive functioning (Field et al., 2024). Terror management theory also suggests this might allow a reinforcement of our worldviews, be they right or wrong, to psychologically shield ourselves from the fear associated with climate uncertainty (Solomon et al, 2015).

So, yes, we are anxious, depressed, angry, and grief stricken. And no wonder! We are confronting what it means to live through a planetary transformation that is so vast compared to what our psyches have faced before. That is cause for a little unraveling. But maybe we are also unraveling *what no longer serves us*. Individualism, loneliness, and isolation—all aspects of a neoliberal order that has compounded mental health issues even prior to climate change showing its effects (Crouch, 2011; Hall & Briskman, 2017). Goelitz (2024) refers to this phenomenon as *shared trauma*—a reality in which the social worker is no longer a distant observer or external support figure, but instead a co-experiencer of collective crisis. In the context of climate change and ecological collapse, this means we share not only empathy with clients but also the actual lived conditions of suffering: the heat, the smoke, the floods, the resource scarcity, and the grief over disappearing futures. While the impacts remain unequal, shaped by privilege, geography, race, and wealth, the *presence* of trauma is global, cutting across species lines, dissolving the old myth that the helper is somehow exempt from harm.

In practice, this means that when a flood hits a town, the social worker may be evacuating their own family while coordinating a disaster response hotline. When a heatwave causes power outages, they may be simultaneously trying to reach vulnerable clients and manage their own health risks. The boundary between professional and personal becomes fluid. Goelitz (2024) rightly warns that this blurring of boundaries could threaten the mental health and effectiveness of social workers if left unacknowledged. She encourages workers to name the shared trauma, hold space to talk about it, and develop practices that allow social workers to continue without breaking down.

However, I would argue that while this is a pragmatic and appropriate framework for single or time-limited disasters, it begins to unravel in the face of a permanent state of crisis. With climate breakdown, we are not dealing with isolated traumatic events but with the ongoing deterioration of the life systems that underpin all forms of stability. There is no post-trauma, only continued adaptation to worsening conditions. Over time, this persistent erosion will not just blur the boundary between helper and helped-it will erase it. As material, psychological, and social resources are stripped one by one, all of us are gradually reduced to our raw shared humanity.

In this context, the social work role in mental health care begins to change. Traditional models of intervention, diagnosis, and structured support lose relevance when systems themselves are collapsing. Instead, what emerges is a shoulder-to-shoulder model: not top-down, but alongside -one human being walking with another. The power differential is flattened not by intention, but by necessity. When the worker no longer has reliable access to food, clean water, safe housing, or systemic backup, they cannot be positioned as savior, expert, or distant protector. They become, instead, a companion in suffering and meaning-making.

The ability to recognize trauma, to translate emotional distress into connection, and to offer even the smallest tools for coping-these remain powerful gifts even in degenerating systems. It asks the social worker to lead not from expertise alone, but from vulnerability, humility, and courage. You can be the one who says, "Yes, I feel it too. And yes, we will face it together."

Tools and Frameworks for Managing Burnout, Eco-Anxiety, and Climate Grief

As climate collapse reshapes the emotional landscape of both clients and practitioners, social workers need concrete tools to navigate burnout, eco-anxiety, and climate grief. This section introduces a set of adaptable tools and frameworks for individual and collective use. They are grounded in trauma-informed practice, emotional regulation, and principles of community care.

Method 1: The Anchor-Map Model

What It is:

A reflective tool designed to help social workers (or clients) identify their emotional anchors, sources of overwhelm, and guiding values or goals. (Similar to eco map values mapping—(Hartman, 1995).)

How It Works:

By mapping out three key categories-anchors (stabilizing resources), stressors (climate-related or systemic pressures), and map points (values or direction)—practitioners can gain clarity on what supports and challenges them in their work.

Why It's Useful:

In chaotic or emotionally overwhelming environments, it's easy to lose track of what keeps us grounded or why we continue doing the work. This model externalizes internal experiences and helps build a foundation for reflection, boundaries, and resilience.

Scenario:

Maria, a child protection social worker in a drought-prone rural area, is experiencing insomnia and persistent anxiety as rising temperatures force more families to migrate, impacting her caseload. She feels overwhelmed but guilty for even feeling that way.

Exercise:

Draw three columns labeled Anchor / Stressor / Map Point. Reflect or write:

- **Anchor**—What helps you feel grounded right now? (e.g., walking the dog, a photo of my grandmother, colleagues who understand)
- **Stressor**—What climate-related experiences or thoughts are emotionally overwhelming? (e.g., feeling like no matter what I do, families will still be displaced)
- **Map Point**—What personal value or direction keeps you going? (e.g., I believe helping is valuable, I want to be someone who shows up)

Reflection Questions:

- What patterns do you notice in your anchors and stressors?
- Are your map points realistic right now? Can they be simplified?

Method 2: Regulate–Relate–Respond Framework

What It is:

A nervous system-informed approach that structures how to respond to stress by first focusing on regulation, then connection, and finally action (Similar to Bruce Perry's 3Rs—(Perry & Pollard, 1998)).

How It Works:

The sequence encourages individuals to first calm physiological stress responses (regulate), connect to others or grounding practices (relate), and only then make decisions or engage further (respond).

Why It's Useful:

Climate-related work often triggers a sense of urgency or emotional overload. This method slows the process down, preventing burnout and reactive patterns, while increasing the capacity for thoughtful, embodied engagement.

Purpose: To reduce burnout by helping social workers regulate their nervous systems before engaging in emotionally demanding work.

Scenario:

Jonas works in a shelter affected by recent flooding. He's short-tempered with colleagues and feels numb around clients, then guilty afterward.

Exercise:

Before starting a shift or after an intense client meeting, go through this 3-step process:

- **Regulate**—Do a 2-min grounding practice (e.g., inhale for 4, hold for 4, exhale for 4, hold for 4; repeat).
- **Relate**—Text a trusted colleague a check-in emoji or message about how you are feeling.
- **Respond**—Reassess the situation or task. Ask: What needs doing now and what can wait?

Reflection Questions:

- What signs tell me I need to regulate before I relate or respond?
- Where can I build "micro-regulation" into my day?

Method 3: Climate Grief Naming Circles

What It is:

A structured, communal process to acknowledge and express grief related to ecological loss and climate collapse. (Inspired by Climate Grief Circles—(GreenFaith, 2023).)

How It Works:

Through metaphor, symbols, and shared space, individuals name what they are grieving—loss of landscapes, species, stability, or imagined futures—within a supportive group.

Why It's Useful:

Climate grief is often unspoken, pathologised, or dismissed. This method honors it as a real and valid emotional response. By grieving together, individuals can reduce isolation and restore meaning and a form of community.

Scenario:

A social work team in a coastal city affected by sea level rise meets monthly for peer support. Several workers mention feeling despair about repeating evacuations and losing places they love.

Exercise (Group or Individual):

1. **Opening prompt**—"If your climate grief today was a landscape, what would it be?"
2. **Naming**—Each person names something they feel grief for (e.g., lost traditions, coral reefs, trust in the future).
3. **Witnessing**—Place a symbolic object (stone, leaf, shell) in a shared bowl as each is named.
4. **Closing**—Group breath or moment of silence.

Reflection Questions:

- What kind of climate grief do I carry: personal, fearing the future, or cultural?
- What helps me feel seen in that grief?

Method 4: The Burnout Compass

What It is:

A simple tracking tool that helps individuals assess their current state across four domains: body, mind, emotion, and purpose (Similar to Burnout self-check—(Thomas Walker consulting, n.d)).

How It Works:

Each domain is examined weekly or regularly, with participants identifying early warning signs of burnout and making small, practical adjustments to restore balance.

Why It's Useful:
Burnout doesn't arrive all at once-it creeps in. This method helps social workers stay ahead of full collapse by making micro-interventions before exhaustion becomes entrenched.

Scenario:

Sara, a school social worker, notices she's eating irregularly, dreading mornings, and having trouble concentrating. Her region has been under constant wildfire threat.

Exercise (Weekly Check-In):

Fill in the following chart:

Domain	Current Sign	One Small Adjustment
Body	Tension in jaw, skipping meals	Prep snacks; schedule stretch breaks
Mind	Trouble focusing, forgetfulness	Use timer blocks with 5-min resets
Emotion	Irritable, emotionally flat	Journal 5 words about my mood each evening

(continued)

(continued)

Domain	Current Sign	One Small Adjustment
Purpose	Questioning if it makes a difference	Re-read feedback from a student I helped last year

Reflection Questions:

- Which domain is calling for the most attention this week?
- Am I confusing exhaustion with failure?

Method 5: Collective Care Agreements

What It is:

A relational and cultural practice that helps social work teams create shared expectations, support structures, and emotional safety in their work together. (Using the idea of collective care—(Social Development Direct, 2024).)

How It Works:

Teams co-create a set of statements or agreements that define how they support each other emotionally, how they share responsibilities, and how they navigate shared trauma or stress.

Why It's Useful:

Individual self-care is not enough in the face of systemic collapse or burnout. This method emphasizes that care must be relational, mutual, and structured at the group level-not left solely to personal responsibility.

Scenario:

A team in a disaster relief agency is facing secondary trauma and turnover. They decide to build a collective care agreement into their weekly meetings.

Exercise (Team Activity):

1. Brainstorm statements such as:

    "We assume good intent, even when we mess up."
    "We commit to checking in emotionally, not just administratively."
    "We rotate care roles: someone always holds space."

2. Choose 3–5 to formalize as a living document. Post it in a shared space.

Reflection Questions:

- When was the last time I let someone support me at work?
- What care practice do I want to normalize in my team?

Conclusion

These tools and exercises are not designed to erase distress or make social workers immune to the climate crisis. Rather, they offer structures of care and reflection that help make overwhelming realities more bearable—and most significantly, shareable. They affirm that we do not need to be emotionally separate from the world to be effective in it. In fact, connection may be our most powerful tool of all.

References

Albrecht, G. (2005). Solastalgia: A new concept in human health and identity. *Philosophy, Activism, Nature, 3*, 41–55.

Berry, H. L., Bowen, K., & Kjellstrom, T. (2010). Climate change and mental health: A causal pathways framework. *International Journal of Public Health, 55*(2), 123–132.

Berry, H. L., Waite, T. D., Dear, K. B. G., Capon, A., Murray, V., & Patchett, G. (2018). The case for systems thinking about climate change and mental health. *Nature Climate Change, 8*, 282–290. https://doi.org/10.1038/s41558-018-0092-2

Brown, H. J. Jr. (n.d.). In the confrontation between the stream and the rock… [Inspirational quote]. In *Life's little instruction book* (2nd ed., p. 45).

Burke, M., González, F., Baylis, P., Heft-Neal, S., Baysan, C., Basu, S., & Hsiang, S. (2018). Higher temperatures increase suicide rates in the United States and Mexico. *Nature Climate Change, 8*(8), 723–729.

Chancel, L., Bothe, P., & Voituriez, T. (2023). *Climate inequality report 2023* (World Inequality Lab Study 2023/1). World Inequality Lab. https://wid.world/www-site/uploads/2023/01/CBV 2023-ClimateInequalityReport-1.pdf

Crouch, C. (2011). *The strange non-death of neoliberalism*. Polity Press.

Doherty, T. J., & Clayton, S. (2011). The psychological impacts of global climate change. *American Psychologist, 66*(4), 265–276. https://doi.org/10.1037/a0023141

Field, J., Jones, B., & Wigner, E. (2024). A theoretical model of climate anxiety and coping. *Discover Psychology, 4*, Article 94. https://doi.org/10.1007/s44202-024-00212-8

Goelitz, A. (2024). *Shared trauma, shared resilience: Building social work practice for a changing climate*. Oxford University Press.

GreenFaith. (2023, August). *How to start a GreenFaith Circle: Circle guide—building a safe community space* [PDF guide]. https://www.greenfaith.org/wp-content/uploads/2023/08/Circle-Guide_-How-to-Start-A-GreenFaith-Circle.pdf

Hall, S. M., & Briskman, L. (2017). Neoliberalism and mental health: An analysis of the individualisation of mental health problems in Australian social policy. *Australian & New Zealand Journal of Psychiatry, 51*(7), 606–613. https://doi.org/10.1177/0004867417703630

Hartman, A. (1995). Diagrammatic assessment of family relationships: The family ecomap. *Families in Society, 76*, 111–122.

Hickman, C., Marks, E., Pihkala, P., Clayton, S., Lewandowski, E., Mayall, E., & Mellor, C. (2021). Climate anxiety in children and young people and their beliefs about government responses to climate change: A global survey. *The Lancet Planetary Health, 5*(12), e863–e873. https://doi.org/10.1016/S2542-5196(21)00278-3

Mulambo, C. H., Thakur, R., & Mathew, S. (2025). Hot weather and violence against women: A global scoping review. *International Journal of Environmental Research and Public Health, 22*(7), 1069. https://doi.org/10.3390/ijerph22071069

Obradovich, N., Migliorini, R., & Rahwan, I. (2024). Ambient heat exposure and sleep disturbances: A systematic review in the context of climate change. *Environmental Health Perspectives*. https://doi.org/10.1289/EHP12345

Oxfam International. (2023, November). *Climate equality: A planet for the 99%*. Oxfam GB for Oxfam International. https://www.oxfam.org/en/research/climate-equality-planet-for-the-99

Perry, B. D., & Pollard, R. (1998). Homeostasis, stress, trauma, and adaptation: A neurodevelopmental view of childhood trauma. *Child and Adolescent Psychiatric Clinics of North America, 7*(1), 33–51. https://doi.org/10.1016/S1056-4993(18)30258-X

Social Development Direct. (2024, February 27). *Tip sheet: Collective care*. https://www.sddirect.org.uk/resource/tip-sheet-collective-care

Solomon, S., Greenberg, J., & Pyszczynski, T. (2015).*The worm at the core: On the role of death in life* (25th anniversary ed.). Random House.

Stanley, S. K., Hogg, T. L., Leviston, Z., & Walker, I. (2021). *From anger to action: Differential impacts of eco-anxiety, eco-depression, and eco-anger on climate action and wellbeing. Journal of Climate Change and Health*, 1, Article 100003. https://doi.org/10.1016/j.joclim.2021.100003

Thomas Walker Consulti (n.d.). *Burnout Self-Check: Where's your energy going?* Retrieved [November 2nd, 2025], from https://www.thomaswalkerconsulting.com.au/Resources-Pages/Burnout-Self-Check-Tool.html

White, B. P., et al. (2023). *Mental health impacts of climate change among vulnerable populations: A systematic review. International Journal / PMC (2023).*

World Health Organization. (2023). *Climate change and health (fact sheet)*. WHO.

Chapter 9
Practical Toolkit for Social Workers

Introduction

The following toolkit is taken from a wide variety of sources, disaster experts, and adaptation planning from various countries' existing disaster plans. As much as possible, I have attempted to provide examples of globally available resources, but there should be more specific and detailed available information in your own country's government websites. I would recommend printing out hard copies of those website resources that you find most helpful and keeping them in a folder in an accessible location. Power supplies may be lost, including internet access so physical copies of useful information will be one less thing to worry about.

Wildfire Resources

Responding to Wildfire: Social Work Action

As a social worker responding to a wildfire, your role can extend beyond direct support to victims—it can include preparedness, crisis navigation, and emotional containment. In communities at risk of wildfire, which is becoming larger than previously experienced, social workers should ensure both they and their clients are ready to act quickly. This includes helping individuals develop evacuation plans, prepare emergency kits, protect essential documents, and understand local emergency protocols. During a wildfire, social workers may also serve as key communication points, ensuring clients receive accurate and timely alerts, especially those with limited digital access. Additionally, they play a critical role in providing emotional support, assisting with shelter access, and helping individuals manage the trauma and disruption wildfires can cause. Empowering clients with emergency planning skills and practical resources could help reduce immediate risk and build a sense of resilience.

S. Rost, *Climate Resilience Guide for Social Workers*,
SpringerBriefs in Social Work, https://doi.org/10.1007/978-3-032-24604-2_9

Social Workers Can:

- Help clients develop and practice evacuation plans, including multiple routes and safe destinations.
- Assist in preparing both Go-Kits (3-day portable emergency supplies) and Stay-at-Home Kits (2-week supply), tailored to the clients' needs
- Encourage clients to keep essential documents (ID, health records, financial info) safe and accessible.
- Support access to a 1-month supply of medications stored securely and accessibly.
- Educate clients on how to sign up for local emergency alerts and stay informed via battery-powered radios or backup phone chargers.
- Check on clients with limited mobility or no transport, helping them plan evacuation options.
- Help individuals prepare for utility failures by planning for medical equipment and power needs.
- Provide emotional and psychological support during evacuation and post-fire recovery.
- Offer to refer to training in basic first aid and CPR, especially for clients in isolated areas.
- Coordinate with local agencies to locate shelters, services, and post-disaster support.

Information drawn and adapted from the Red Cross wildfire checklist (American Red Cross, n.d.).

Heat Stress Resources

Social workers can play a role in both preparing communities and intervening during periods of extreme heat. Practitioners can support clients holistically by combining medical knowledge with social and emotional care. This includes assessing vulnerability, promoting prevention strategies, and providing direct assistance as temperatures rise. Key responsibilities include checking on isolated or high-risk individuals, coordinating community cooling strategies, and offering emotional support to manage the stress of sustained heat exposure.

Social Workers Can:

- Identify and check on vulnerable individuals such as older adults, young children, people with chronic illnesses, and those living alone.
- Promote heat illness recognition and first aid by educating clients on symptoms of heat exhaustion and heatstroke.
- Help clients prepare cooling strategies such as closing curtains, wearing loose clothing, staying hydrated, and accessing cool spaces.
- Support emergency readiness by encouraging the use of heat alerts, battery-powered radios, and backup phone chargers.

- Coordinate community cooling actions by collaborating with local shelters, libraries, community centers, and mutual aid groups.
- Advocate for systemic adaptations such as neighborhood buddy checks, community cooling protocols, and heat-resilient home upgrades.

Adapted from information provided by the National Health Service, United Kingdom (NHS, 2022).

Water Shortage Resources

Social Work Response to Emergency Water Shortages

In the face of a sudden water shortage, whether due to drought or infrastructure failure, social workers may become crucial actors in helping affected individuals navigate the crisis. Their responsibilities include identifying vulnerable clients, ensuring they can access emergency water supplies, and providing accurate, up-to-date information on local water distribution efforts. Social workers also help clients maintain hygiene and hydration under constrained conditions and support emotional wellbeing as people deal with fear and uncertainty. Acting as a bridge between community needs and local authority response, social workers can help prevent the most at-risk populations from being left behind. As climate impacts increase, social workers may also be required to undertake practical water harvesting education and provide resources to do this.

Social Workers Can:

- Identify and map vulnerable populations in advance—including older adults, people with disabilities, young children, and those without reliable transport or digital access—so they are included in future emergency water planning.
- Educate communities proactively about how to safely collect, store, and ration water during shortages, incorporating culturally relevant and accessible materials.
- Support clients in adapting homes for future scarcity, such as by installing rainwater collection systems, greywater reuse, or low-water hygiene alternatives.
- Facilitate community-level water-sharing agreements or cooperatives, ensuring social equity and avoiding conflict in resource-limited settings.
- Build local knowledge of water purification methods, including boiling, filtration, chemical treatment, and solar disinfection-particularly in areas likely to experience contamination from drought or flooding.
- Develop emergency water access plans for clients with medical dependencies (e.g., those needing hydration for medications, dialysis, or cooling for heat-sensitive conditions).

Information adapted from Sweden's risk advisory groups MSB (MSB, n.d.) to project future needs.

Flash Flooding

Floods can escalate quickly and become life-threatening, especially flash floods which come quickly with vast amounts of water. Social workers should be prepared to protect themselves, assist others, and communicate clearly about safety procedures. Alongside emotional support and coordination, part of your role could be to help clients understand how to reduce their physical risks.

Social Workers Can Advise Clients:

Before a Flash Flood or Storm Surge

- Emphasize urgency: Flash floods can occur within minutes or hours. Stress to clients that early action saves lives-waiting is not an option.
- Know and share high-risk zones: Make sure clients understand whether they live in a low-lying, coastal, or urban drainage area especially vulnerable to rapid flooding.
- Identify vertical evacuation routes: Help clients locate nearby higher floors, rooftops, or hills that are reachable on foot in case horizontal evacuation is impossible.
- Encourage grab-and-go kits: Instruct clients to keep emergency kits packed and near exits. Include water, flashlight, medication, cash, documents, and clothes in a waterproof bag.

During a Flash Flood or Storm Surge

- Get to high ground immediately: Emphasize that climbing to safety is better than trying to drive or run through floodwater.
- Never walk or drive through moving water: Just 15 cm (6 in) of water can knock a person down, and 30 cm (12 in) can float a car. Strong currents are often invisible.
- Evacuate vertically if horizontal evacuation is blocked: Encourage clients to go to the highest safe level of a sturdy building and signal for help.
- Avoid basements, underground spaces, or storm drains: These fill quickly and can trap people with little warning.
- Stay off bridges and coastal areas: Storm surges can destroy infrastructure rapidly and create dangerous waves and debris.

Information gathered and adapted from the Federal Emergency Management Agency (2022). *Flood safety tips.* Ready.gov. https://www.ready.gov/floods.

Storm Disaster Resources

Social workers are essential in helping vulnerable individuals prepare for and respond to severe weather events such as storms, high winds, hurricanes, and cyclones. Staying safe yourself ensures you can support others effectively. Preparation includes helping clients secure their homes, plan safe movements, and access emergency information. When a storm hits, it's important to prioritize personal safety, maintain communication, and provide both practical aid and emotional reassurance. After the storm, social workers play a key role in assessing damage, connecting clients with recovery resources, and addressing stress or trauma.

Social workers can advise or directly help with:

- Helping clients secure loose outdoor items (e.g., garden furniture, bins, trampolines) to reduce risk from high winds.
- Encourage signing up for local weather alerts and checking official guidance regularly.
- Help clients prepare emergency kits (torch, batteries, first aid, radio, flashlight, contact list), and ensure they have alternative power sources and document backups.
- Advise clients to stay indoors, away from windows and exterior walls, and avoid going outside to fix damage.
- Remind clients not to drive through floodwater; always turn around if water covers the road.
- Proactively check in on vulnerable clients such as older adults or those living alone, before and during the storm.
- Coordinate with local agencies to connect clients with clean-up support, temporary shelter, or financial aid.

Information gathered and adapted from Met Office UK (Met Office, n.d.).

Food Security

Social workers can actively support food security during climate impacts by helping communities develop local and resilient food systems. This includes supporting the creation of food kitchens, food hubs, community gardens, and food-sharing initiatives. These grassroots efforts not only reduce hunger but also foster social cohesion, mutual aid, and long-term adaptation. Social workers can assist with identifying local leaders, organizing volunteers, connecting people with land or funding, and advocating for local policies that support community-based food systems. Engaging communities in food growing and sharing increases resilience and ensures equitable access to food in times of crisis.

Resources for Community Food Initiatives

Sustain—Setting up a Community Food Project

Guidance on how to establish food co-ops and community food projects, including planning and funding advice.

Link: https://www.sustainweb.org/foodcoops/how_to_set_up_a_food_co_op/.

Incredible Edible Network

A UK-wide network supporting communities to grow food in public spaces and build food resilience.

Link: https://www.incredibleedible.org.uk/

Food not Bombs—Guide to Starting a Food-Sharing Group

A practical guide for starting a local food-sharing group using surplus food and volunteer networks.

Link: https://foodnotbombs.net/new_site/start_a_group.php

FAO—Urban Agriculture Toolkit

Toolkit from the UN Food and Agriculture Organization for supporting urban agriculture and community gardens.

Link: https://www.fao.org/3/ca2527en/CA2527EN.pdf

Transition Towns—Local Food Systems

Resources to build local food systems as part of wider community resilience strategies.

Link: https://transitionnetwork.org/do-transition/starting-transition/community-food/

Migration

Climate change is already contributing to increased migration as people are forced to leave areas affected by rising seas, extreme heat, drought, or repeated disasters. In the future, social workers may increasingly support climate-displaced individuals and families, both within and across borders. This could involve helping with access to housing, healthcare, legal aid, and emotional support, while also navigating cultural, language, and trauma-related challenges. Social work will need to adapt by becoming more informed about climate-related drivers of migration, advocating for rights-based responses, and building community systems that can respond with compassion, inclusion, and practical resources for those uprooted by environmental change.

Tools for Social Workers Supporting Climate Migrants and Refugees

Holistic Needs Assessment Template

Covers housing, legal status, trauma, medical needs, and social supports.

Link: https://www.refworld.org/docid/5f2082b94.html

Psychosocial First Aid (PFA) Guide

A step-by-step method to provide calm, supportive care in crisis or displacement situations.

Link: https://www.who.int/publications/i/item/9789241548205

Migration Story Mapping Tool

A simple timeline or drawing exercise to help clients process and narrate their migration experience.

Link: https://www.rwint.org/toolkit/narrative-mapping

Cultural Formulation Interview (CFI)

A structured way to understand a person's background, cultural identity, and how they experience stress or illness.

Link: https://www.psychiatry.org/psychiatrists/cultural-competency/cultural-formulation

Rights Awareness Leaflet (Multilingual)

A handout summarizing basic rights to housing, healthcare, education, and safety—adapted for people without formal refugee status.

Link: https://picum.org/resources/

Community Resource Map Template.

Helps clients identify key services, safe places, and support networks in their new environment.

Link: https://www.redcross.org/get-help/how-to-prepare-for-emergencies/make-a-plan.html

Peer Support Group Framework

A guide for setting up culturally sensitive, trauma-aware support groups for climate-affected migrants.

Link: https://www.unhcr.org/uk/protection/health/5f9a7cae4/community-based-mental-health-and-psychosocial-support-approaches.html

Children and Youth Resilience Tools

Social workers play a vital role in supporting children and young people affected by the physical and emotional impacts of climate change. These impacts may include displacement, trauma from extreme weather events, eco-anxiety, grief, and instability in their communities or families. Social workers can provide safe spaces for expression, validate their concerns, support families, build resilience through community engagement, and offer practical coping tools tailored to children's developmental stages. Collaborating with schools, parents, and health services ensures children's voices are heard and their rights protected in a climate-impacted world. Having an awareness of the increased risks children face as a result of climate changes such as violence, respiratory difficulties, and heat stress is important in risk assessment considerations.

Resources for Social Workers Supporting Children and Youth

UNICEF—Climate Crisis is a Child Rights Crisis

Global report and tools emphasizing the vulnerability of children to climate change and actions for child-focused climate resilience.

Link: https://www.unicef.org/reports/climate-crisis-child-rights-crisis

Save the Children—Climate Change Resources

Articles, advocacy tools, and reports on how climate change affects children and what can be done to protect them.

Link: https://www.savethechildren.org.uk/what-we-do/climate-change

Child and Youth Resilience Measure (CYRM)

A validated tool to assess resilience in children and youth, useful for social workers in post-disaster or adaptation contexts.

Link: https://cyrm.resilienceresearch.org/

Talking with Children About Climate Change—Child Mind Institute

Practical guide for helping children manage fear and anxiety related to the climate crisis.

Link: https://childmind.org/article/talking-to-kids-about-climate-change/

Climate Psychology Alliance—Resources for Young People

Supportive materials for engaging young people in conversations about climate emotions and resilience.

Link: https://www.climatepsychologyalliance.org/index.php/resources/young-people

Australian Psychological Society—Guide for Children and Climate

Child-specific advice and emotional support strategies for managing climate-related stress and trauma.

Link: https://psychology.org.au/for-the-public/psychology-topics/climate-change-and-mental-health

Community Support and Mutual Aid Resources

Social workers are well-equipped to initiate and sustain mutual aid networks in communities affected by climate-related events. In practice, this can involve coordinating neighborhood check-in systems for vulnerable residents, such as elderly people or those without transportation, organizing community food shares or water distribution hubs, and setting up support groups for those experiencing climate grief or displacement. Social workers can use their skills in facilitation and advocacy to bring together local residents, faith groups, schools, and NGOs to co-develop shared emergency response plans. They also help ensure inclusive participation, such as providing interpretation for non-native speakers, making meetings accessible for disabled residents, and ensuring child care so caregivers can attend. By mapping local assets (skills, spaces, supplies) and creating communication pathways (e.g., WhatsApp groups, support networks of volunteers, email contact), social workers strengthen community self-reliance. Importantly, they can also navigate bureaucracy, connecting grassroots groups with emergency services, housing agencies, or public health resources.

Resources for Building Mutual Aid and Community Support Networks

Big Door Brigade—How to Start a Mutual Aid Network

Comprehensive guide for forming mutual aid groups, including principles, practical steps, and examples.

Link: https://www.bigdoorbrigade.com/mutual-aid/

Mutual Aid Disaster Relief—Organizing Tools

Tools and training for grassroots disaster response and mutual aid efforts rooted in solidarity, not charity.

Link: https://mutualaiddisasterrelief.org/

UK Mutual Aid Resources and Networks

Directory and insights into mutual aid organizing in the UK context, useful for community building during crises.

Link: https://freedomnews.org.uk/mutual-aid-in-the-uk/

Transition Network—Building Local Resilience

Global movement supporting communities to become more resilient through practical projects, networks, and training.

Link: https://transitionnetwork.org/

Cooperation Humboldt Mutual Aid Toolkit

Toolkit with planning tools, facilitation guides, and community mapping strategies.

Link: https://cooperationhumboldt.com/mutualaid/

Mental health support

As climate impacts worsen, social workers can provide vital mental health support by focusing on practical, realistic steps. In emergencies, one key role is checking in on clients who rely on mental health medication or addiction support, helping them navigate disruptions if clinics are closed or supplies run out. This might involve contacting local pharmacies, connecting with emergency services, or helping clients identify temporary alternatives. Social workers can also offer basic emotional support-listening, validating distress, and helping people feel less alone. Where possible, they can guide clients toward existing mental health services, online supports, or community-based peer networks. Rather than taking on everything, the focus is on doing what's possible with the resources at hand: helping people stay stable, feel seen, and know where to turn next.

Mental Health Support Resources

- **Climate Mental Health Network**
  Resources, toolkits, practitioner directory, and community events
  Link: https://www.climatementalhealth.net/resources
- **Climate Psychology Alliance**
  Articles, webinars, support spaces, and a therapist directory
  Link: https://www.climatepsychologyalliance.org
- **Good Grief Network**
  Peer-to-peer support groups for climate grief and anxiety

Link: https://www.goodgriefnetwork.org

- **MHCCA—Mental Health and Climate Change Canada**
  Self-assessments, climate coping tools, and educational resources
  Link: https://mhcca.ca
- **Force of Nature**
  Youth-focused climate anxiety workshops and climate cafés
  Link: https://www.forceofnature.xyz
- **Climate Anxiety Counseling (Climate Clock)**
  Mental health tips, hotline connections, and group events
  Link: https://climateclock.world/mental-health-awareness
- **Gen Dread by Britt Wray**
  A newsletter and archive on eco-anxiety, activism, and resilience
  Link: https://gendread.substack.com
- **Commons Social Change Library: Climate Grief Resources**
  Practical guides for coping with eco-grief and climate distress
  Link: https://commonslibrary.org/resources-to-cope-with-climate-anxiety-and-grief/
- **Cornell Health—Climate and Eco-Anxiety**
  University health resource page with coping strategies
  Link: https://health.cornell.edu/about/news/eco-anxiety

Reflection Exercise

Scenario 1: The Flooded Town

Context:

After two weeks of extreme rainfall, a small town has experienced severe flooding. Many homes have been damaged or lost entirely, and several neighborhoods are without power or clean drinking water. You are a social worker in the area and are tasked with helping coordinate support for displaced families, many of whom are already facing economic hardship.

Key Details:

- Several of your clients are among the displaced.
- One client with chronic health conditions has lost their medication and mobility aids.
- The emergency shelter is full and under-resourced.
- You, personally, are also affected—your route to work has been cut off and your own home has minor damage.

Reflection Questions:

1. What are the immediate social work priorities in this situation?
2. How do you balance your role as both an affected person and professional?
3. What mental health needs might arise now and in the weeks that follow?

4. What resources from this toolkit could support both your clients and your own wellbeing?
5. How can you engage collective care or mutual aid strategies to reduce overwhelm?

General Question:

How can you, and your team if you have one, map the vulnerability of your client group? How might this help in preparations for future climate impacts?

References

American Red Cross. (n.d.). *Wildfire safety checklist*. https://www.redcross.org/content/dam/redcross/get-help/pdfs/wildfire/EN_Wildfire-Safety-Checklist.pdf

Federal Emergency Management Agency. (2022). *Flood safety tips*. Ready.gov. https://www.ready.gov/floods

Met Office. (n.d.). *Stay safe in a storm*. https://weather.metoffice.gov.uk/warnings-and-advice/seasonal-advice/your-home/stay-safe-in-a-storm

Myndigheten för samhällsskydd och beredskap. (n.d.). *Water shortage and drought*. Krisinformation.se. https://www.krisinformation.se/en/hazards-and-risks/water-shortage-and-drought2

National Health Service. (2022, August 10). *Heatwave: How to cope in hot weather*. NHS. https://www.nhs.uk/live-well/seasonal-health/heatwave-how-to-cope-in-hot-weather/

Chapter 10
Conclusion: Holding the Light in a Darkening World

The times we live in are precarious and unprecedented. These are not phrases we ever wanted to confront in our lifetime, let alone integrate into our daily work. Yet here we are, in a moment that is reshaping everything we know about safety, certainty, and systems of support. Climate change is no longer a distant threat. It is a lived, uneven, escalating reality. And as social workers, we find ourselves on the frontlines of its human consequences.

We do not want this burden. We didn't ask for it. But it is ours to carry, whether we choose to look it in the eye or not. And many days, it may feel too heavy. Overwhelming. Crushing. There will be moments of grief, paralysis, anger, denial, and despair. We may ask ourselves, *What should I do? What can I possibly do that will matter?*

In this way, climate collapse is not only a physical threat but a deeply psychological and spiritual one. It touches the oldest parts of our minds-those tuned to survival and fear of loss. It may drive us into numbness or cynicism or urge us to retreat into personal safety and routine. It may tempt us to look away.

But there is a truth that lives beyond the fear: if you feel alone, you are not alone in that feeling. And that shared sense of being adrift in uncertainty-that shared ache-is exactly where our collective strength begins.

When people hear hard truths about the future, they often ask a very human question: *What now?* It's the same question you might ask if you were given a terminal diagnosis or knew that time was running out. *What still matters? What gives life meaning?*

And here lies one of the most powerful insights of all that I have discovered: social work still matters and will matter even more as time goes on.

In a world where the fabric of everyday life is fraying, the simple act of care is powerful. Choosing to help, to listen, to share, to organize, and to build community-these are not just survival strategies. They are acts of resistance against despair. Social work is not just a profession in this moment. It is a deeply human response to chaos. It is a container for meaning.

S. Rost, *Climate Resilience Guide for Social Workers*,
SpringerBriefs in Social Work, https://doi.org/10.1007/978-3-032-24604-2_10

In the face of collapse, we can choose to be part of something that still builds rather than breaks, that still connects rather than isolates. Without trying to be too saccharine-filled, or overestimate what social workers can do—I sincerely believe that we can be a flicker of light in dark times that gives hope to those in pain, that we can remind people: *we are in this together.*

This is not a call to heroism. It is a call to honesty. We will not save the world. But we may help each other live in it. We may help make it more bearable, more just, and more compassionate—for however long we can.

And when things become too much, we can remember: it is not only okay to rest—it is necessary. Our ability to keep going is not going to be powered by martyrdom but by connection and meaning. We are not meant to hold this alone.

Let this work remind you of who you are. Not a savior. But a person—doing the slow, sacred work of showing up.

And if you are showing up, holding even the smallest light in the darkness, I thank you and please remember that light matters more than you know.

Index

S. Rost, *Climate Resilience Guide for Social Workers*, SpringerBriefs in Social Work, https://doi.org/10.1007/978-3-032-24604-2

MIX
Papier aus verantwortungsvollen Quellen
Paper from responsible sources
FSC® C105338

If you have any concerns about our products,
you can contact us on
ProductSafety@springernature.com

In case Publisher is established outside the EU,
the EU authorized representative is:
Springer Nature Customer Service Center GmbH
Europaplatz 3, 69115 Heidelberg, Germany

Printed by Libri Plureos GmbH
in Hamburg, Germany